T0115360

In the Little World

(A True Story of Dwarfs, Love, and Trouble)

JOHN H. RICHARDSON

An *Abacus* Book

First published in the United States of America in 2001
by HarperCollins Publishers Inc.
First published in Great Britain in 2002
as an Abacus paperback original

A CIP catalogue record for this book
is available from the British Library.

ISBN 0 349 11497 8

Printed and bound in Great Britain
by Clays Ltd, St Ives plc

Abacus
An imprint of
Time Warner Books UK
Brettenham House
Lancaster Place
London WC2E 7EN

www.TimeWarnerBooks.co.uk

For Julia and Rachel, my future.

One

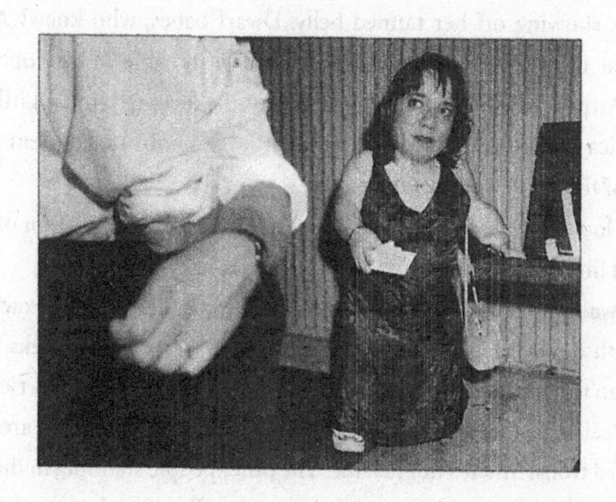

ONE I CAN HANDLE. TWO, NO PROBLEM. But the lobby of this hotel is absolutely *swarming* with dwarfs. There's an old bald dwarf wearing wire-rim glasses, a hipster dwarf sporting a snazzy goatee, Gen-X dwarfs in T-shirts that say NORMAL PEOPLE WORRY ME, a gang of teenage dwarfs cruising by with the insular telepathy of teenagers everywhere—they don't have to look at each other to know they're all there, marching shoulder to shoulder through the clueless adult world. There are dwarfs with normal heads and dwarfs with the classic pushed-in dwarfy look—big forehead, retracted nose, and bulged-out chin, as if God had pushed his thumb on the bridge of their nose and everything else squeezed out accordingly. A few have jumbo-size heads. Some have robin's-breast chests that make them look like tiny superheroes. Others zip around in motorized wheelchairs and scooters, a few lying flat and others kneeling and some scrunched down into their chairs like crabs into shells. And there's one—over by the placard that says WELCOME TO ATLANTA, WHERE LITTLE

PEOPLE ARE SPECIAL PEOPLE—who isn't much more than stick legs jammed into a head. And another one all twisted and folded into himself like a human pocketknife, his muscular arms festooned with tattoos. And let us not overlook the dwarf women, like that classic California sand bunny with the teased blond hair . . . or that Bianca Jagger look-alike . . . or that cute little thing in the tank top, showing off her tanned belly. Dwarf babes, who knew? And who knew that flat-chestedness would be such a rarity, here in the lobby of the Atlanta Airport Marriott Hotel at the start of the fortieth annual Little People of America convention? Their big butts and big boobs make them almost a parody of the *Playboy* ideal. Or a rebuke.

And look at this, here in the hallway leading to the convention office, this innocent little baby riding a tiny tricycle. How sweet, how adora—

But wait. Something is wrong. The baby looks like a baby, exactly like a baby, with a perfect bald baby head and puffy white baby cheeks. But real babies can't maneuver like that, can't frown with that look of worried intelligence. Real babies can barely keep their heads from wobbling around like dashboard trolls. And it's not just me. The other people standing in the hallway are also watching him with the same charmed and confused expression, clearly feeling the same oddly pleasant subliminal dislocation.

And that pretty young woman with the lank hair and exhausted eyes? She's about two feet taller than most of the other people in the hall and she's been watching the not-a-baby like a mama hawk. And now she's zeroed in on me, following my lips as I make verbal notes into my tape recorder, eyes boring away like she's trying to see right through me—right through me into my corrupt and voyeuristic soul. To put us both at ease, I step over and ask if she'd mind answering just a few questions, please.

"I have a lot of questions for *you*," she says. "Where are you from? What are you writing?" She's so upset she's starting to cry.

"I—I'm a reporter," I say. "I'm here to cov—"

"Who said you could be here? What did you put on that tape about my son?"

"I have permission to be here," I say, feeling the eyes of the dwarfs in the hallway upon me. Can they see into me too? See the teenage poseur with the morbid streak and the dog-eared copies of Franz Kafka and Zap Comix; see him toking up for that memorable midnight screening of Todd Browning's *Freaks* on St. Marks Place when everyone chanted along at the wedding banquet of beauty and the dwarf: *we accept her, we accept her, one of us?* Man, how juvenile that seems now. These are real people, not some clever transgressive *ficciones.* Even the most twisted of them zip through the lobby in their custom wheelchairs, looking for a friend. So I talk on, alarmed by her alarm, trying to reassure her, telling her I'm here to write an article for *Esquire* magazine and I talked to Lee Kitchens of the Little People of America and even joined the LPA myself—paid my hundred dollars and joined. But she isn't listening. She's glancing around, grabbing my arm, looking for help. She asks someone to watch her son and tugs on my arm, pulling me down the hall toward the LPA office. "I want to find out what you're doing here," she says, still on the verge of tears.

I tell her it's okay, I'm coming, please don't worry.

"Some people are very exploitative. I want to know what you're doing."

I don't argue with her. She's so upset, it's clear that the last thing she needs is an argument. I want to tell her I'm not some kind of *Hard Copy* scumbag but a genuine sensitive guy and serious writer who just spent a whole nightmare summer helping his mother deal with the cancer behind her nose, a summer of CT scans and MRIs and sonograms and echocardiograms and angiocardiograms and two complete physicals and at least a dozen debriefings of her medical history and hours and hours of waiting in hallways alongside patients slumped in wheelchairs and stretched out on gurneys with their chemo-yellow skin and that disturbing radioactive peach fuzz that clings to their bald heads like some kind of mutant postnuclear algae . . . and she just had surgery and is in the hospital recovering at this very moment and the radiation treatments are next and I just came here for a little friggin' *relief.* I've outgrown the weird science phase, really I have. And I could be in Hollywood interviewing celebrities (as I did for years) but I *chose* this because I wanted to write about humanity and truth and obviously this is not the time or the place and the fact is I can't help feeling this odd dislocation and . . . there are just so *many* of them . . . all so *different.* So I keep my mouth shut and let her pull me along and when we get to the office, she finds the person in charge and holds back her tears so she can tell him the story. "He said that he had talked to Lee, but he doesn't know Lee," she says, still clutching my arm. "I want to know what he's doing here. Does he have permission to be here?"

The lady behind the desk talks in a soothing voice. "He's a reporter," she says. "It's okay."

Then the woman's tears finally break loose. Her cheeks get wet, and through the hard fingers on my arm and the tremble in her voice, I can feel the tension washing out of her. "I just want to know he's going to do a good story," she says.

"He's been credentialed. He's going to do a positive story."

"You should make an announcement," she suggests, her voice still shaky. Then she turns to me. "I owe you a Coke." She laughs and sniffles. "I'm sure this has happened to you before."

"Actually, no," I say. "This is the first time."

She's smiling, but she's still holding my arm, holding it tight. And trembling.

. . .

The first thing they notice is the ramp. There are three steps up and three steps down, and in between a platform that goes all the way across the front desk. Which means that a person three or four feet tall can check into the hotel with dignity, can stand there on the platform and hand over a credit card and sign the register and look the clerk in the eye. Which might seem like a small thing, but not if you've come all the way from Australia, not if you just got off an endless twenty-one-hour flight that cost thousands of dollars you don't have. Under those circumstances, that ramp will seem like a sweet welcoming miracle. And so will seeing the dwarfs walking through the lobby, striding along so confident and strong—and seeing dwarfs at the telephone booths, standing on stools provided by the hotel—and seeing dwarf husbands with dwarf wives and dwarf children in dwarf clothes that don't look homemade. As her daughter stares, Evelyn Powell sneaks a look at her face and feels a rush of hope. They've traveled halfway around the world, left so much grief and fear behind them, and it's only now, after this short walk across the lobby of the Atlanta Airport Marriott Hotel, that they've really begun their journey. Later she records her reaction on her laptop computer: *Watching*

Jocelyn's face was magic—surprise, staring and shyness and excitement all at once.

But they can't *use* the ramp. It's a sweet welcoming miracle and they can't use it because of the very thing that brought them here—the wheelchair that Jocelyn started using at ten years old, the wheelchair that has limited and defined and cursed their lives for six years now. So Evelyn pushes her daughter up to the one check-in station reserved for tall people.

That's okay. They're used to it.

Now it's late and they're too tired to breathe. Evelyn closes her file and puts the laptop on the beside table and turns off the light.

"I'm twitching," Jocelyn says.

Evelyn feels a shot of panic. "Is it a nerve? Is it your spine?"

"No, just twitching. It feels funny."

She turns the light back on and examines the terrible curve of her daughter's back. But there's nothing she can do.

"You should use your wheelchair more often," she says, and turns out the light.

. . .

Out by the shot-put area, six little guys warm up. A few have the washboard bellies and bulging arms of serious bodybuilders, and one looks particularly macho with an American-flag scarf tied around his forehead, but not one comes within six inches of my chin. Kyle starts off the shot put, floating quite professionally onto his toes and pointing with one hand before starting his spin. He's from North Dakota and took third place in the state wrestling championships, competing (he tells me proudly) against "average-size" people.

When the ball lands, the referee calls mark. He's a dwarf too, wearing mirrored sunglasses and a baseball cap and looking just like a small version of a southern sheriff, which would be cute if he wasn't scowling.

Michael's up next. He's an actor from Los Angeles, a serious and handsome little man in a black Nike beret sufficiently media-savvy to spell his name for me without being asked: G-i-l-d-e-n. He had a part in *Pulp Fiction* but wants me to mention he's mostly a stockbroker—actors are suspect here, summoning the memory of Munchkins, Oompa Loompas, and a thousand circus sideshows. Plus he plays an elf in the Radio City Music Hall Christmas Show and he gets a *ton* of shit for that.

Michael frowns hard, cocks the heavy ball against his shoulder and bounces it a few times, testing its heft. Then he twirls and throws. The assistant unspools the tape and calls out, "Six point eight five."

Michael scowls. "Six point eight five? I'm in second place now with a crappy six point eight five? I've got to break seven or I'm going to be unhappy."

As the shot put gives way to the javelin toss, Michael tells me that I have to know a few things if I want to do a good story, and the first is that everything may look casual and relaxed out here on the track but this whole convention is a deep-stress situation and already the stress is starting to build. Dwarfs are rare. For every ten or fifteen thousand tall people, one dwarf is born, which adds up to about twenty-five or thirty thousand dwarfs spread all across the country. So most of them live their whole lives among tall people, rarely ever seeing another dwarf, and this week represents their only real chance to meet anyone. That's why each convention sparks about a dozen marriages and as many divorces, and why he's so tense right now, because he met this girl on the Internet and he's been talking to her for six months and today—at exactly five o'clock this afternoon—he's going to meet her in the flesh for the first time. "I have people say, have a good vacation, and I say, 'I'm not going on vacation.' For certain people at certain times in their life, this is a *pivotal* six days—it could change their lives forever."

A power lifter named Adam edges over, shy but eager to contribute to this important theme. "When you get back home, you're a different person," he says, "but when you're here—"

"There's a magic here. Finally you have an advantage—finally you're not competing with a bunch of tall guys."

"You don't want to blow it either, early in the week," Adam says.

"As the week goes on, it's *very* dramatic. The clock is ticking—it's almost Thursday, it's almost Friday. . . ."

"By Wednesday," Adam says, "it's gone."

. . .

See the group of dwarfs in khakis and plaid shirts, all four feet something with the same bulging foreheads and the same short arms and the same bowed legs.

"That's a pack of achons," Beth Tatman says.

See the very tiny woman in her tiny formal dress.

"She has SED—spondyloepiphyseal dysplasia," Beth says. " 'Spondylo' means that having to do with the spine. 'Epiphyseal' is your growth plates—when they seal off too early it stops growth—and 'displasia' means irregularity."

See the woman in plaid shorts, four feet tall with proportional arms and legs.

"She has a condition called Turner syndrome," Beth says. "Turners are all female. They have either a missing or a partially missing chromosome. Instead of being double X for female, they're X blank or X partial blank, and many times they're female by appearance but they do not develop secondary sexual characteristics without hormone treatment. Most cannot have children. Typically, they have a condition called webbing of the neck, where you see a tightness of the neck, like the cords are standing out."

Beth is giving me a guided tour of dwarfism as it appears here in the Marriott lobby. She's stuck in a wheelchair herself, four feet three inches if she could stand, but she's a feisty thing who teaches a class in dwarf sexuality and brags in a honeyed southern voice about skydiving and traveling around the world and gettin' it on—a sixties kind of gal. When I spot the tricycle baby again, she acts as if she sees babies on tricycles every day. "He looks like cartilage-hair hypoplasia to me. The baldness, that's common with cartilage-hair hypoplasia. If you see men and women with very thin, very blond to platinum type hair, that's what it is."

When a particularly twisted little man shudders by using a pair of those walking sticks that have gray plastic grips cupping each forearm, Beth points out his custom shoes, which are almost perfectly circular. "Club feet," she says. She draws my attention to his hands, which seem to be pointing backward. "That's called a 'trigger thumb' or 'hitchhiker's thumb.' " She tells me that he comes from a family of nine children, five of whom have a genetic dwarfism called diastrophic displasia. In addition to the club feet and trigger thumb, this dwarfism is marked by cauliflower ears and cleft palates. For some reason, it's particularly common in Finland.

Beth quickly clears up some of the taxonomic confusions I've been having, like the difference between dwarfs and midgets. In the old days, she says, people used "dwarf" to describe the classic dwarfy look of stunted limbs and bulging forehead, which is generally the result of genetic defects. "Midget" was used for the "proportional" dwarfs who look like small versions of average people, usually the result of hormonal deficiencies in the pituitary gland. But these days dwarfs frown on such distinctions and despise the M-word: if you're four ten or under, you're a dwarf.

Although Beth is so relentlessly relaxed, I'm finding it uncomfortable reducing people to symptoms and types. Does she assume that's my only interest? So I turn her back to the club-footed man and ask what kind of guy he is as a guy.

"He's a pretty cheerful fellow," she says, "a retired accountant for the state of Connecticut. A bachelor. In fact, his brother tried to hook me up with him. I said, 'How's his heart, 'cause I don't want to kill him?' "

She cackles.

Still, it's unsettling. A little girl rides by in a motorized pink wheelchair and Beth says, "She has Morquio's syndrome. One of the mucopolysaccharidosis group of genetic disorders. Typically they pass away by their late twenties, of respiratory failure."

The law of genetics is so strong! If you have hypochondrogenesis, your head will come in an oval shape, you will have a very small chin, and you will run the risk of fatal respiratory distress. If you have Russell-Silver syndrome, you will have a triangular face, your ears will be unusually low on your head, and your little fingers will curve inward. If you have primordial dwarfism, you will be extremely tiny and have a squeezed "birdlike" face. If you have thanatophoric dwarfism, you will die at birth or soon after. This genetic detailing is why I keep mixing up the pseudoachondroplastics—with their big heads and little bodies and strikingly regular faces they're like human clip art, almost completely interchangeable.

But the dwarfs seem to find their similarities comforting. That's an achon, that's an SED; this is not our fault and these are our people. As she tells me about the recent scientific achievements, Beth seems almost smug. Dwarfs are the most "genetically isolatable" humans on the planet (a nicely distanced scientific way of describing my clip-art problem). Most of the forms of dwarfism were named and analyzed in the fifties and sixties, a push that began under a prominent geneticist named Dr. Victor McKusick at the Johns Hopkins Hospital in Baltimore. They got a little carried away, listing about three hundred separate syndromes at one point, but now it's back down to about two hundred. Achondroplastics are the most common, occurring at the rate of about one for every 25,000 births, followed by SEDs (one in 95,000) and diastrophics (one in 110,000). In 1994, a California geneticist named John Wasmuth discovered the gene for achondroplasia, and since then scientists have narrowed the problem to something called the "fibroblast growth factor receptor." In normal people, this receptor turns on when it's time for adolescents to stop growing, otherwise your bones would just keep growing and growing until gravity (which is what kills giants) strained your organs to failure. The achondroplasia gene switches the receptor on and leaves it on, slowing growth to a crawl. The research has been moving so fast that there will soon be fetal tests for the dominant forms of dwarfism, which means that people could eventually be able to abort dwarfs out of existence—substituting "personal selection" for "natural selection" (as Wasmuth put it, somewhat nervously, after discovering the gene). Couple that with rapid progress in drug treatments for bone growth, and dwarfism may be the first and most vivid form of difference to become optional—and the Little World will end. No wonder, then, that dwarfs have discovered identity politics with a vengeance, and quite a few of them say they'd refuse any magic cure because being short is "who they are." One couple even told their doctor they'd use the fetal test to abort an average-size baby. Beneath this anger is the uncomfortable knowledge that dwarfs will never assimilate. As long as movie stars have full lips and oval faces, as long as women dream of "tall, dark, and handsome," dwarfs are the difference that stays different. No dwarf will ever be able to "pass" for tall. All of which places them—as the dwarf scholar David Berreby put it—"on the cutting edge of the dilemma of science and identity."

Then a small Hispanic man passes by. He doesn't look dwarfy at all, and he's a few inches taller than the four-foot-ten-inch limit. "He's a pituitary—just a short man," Beth says. "And a very bad dresser."

I laugh, but she doesn't. "I know other pituitaries, and they are all very bad dressers. I don't know if there's a genetic link or not, but it seems to run in that type."

She's really not kidding.

. . .

I said I chose this story. But the truth is, I stumbled into it with no thought deeper than "sounds interesting, could be different." Or so I thought when I got on the plane for Atlanta. A few more days pass before I remember that I've written about a dwarf before, the son of a Hollywood actress who raised him in isolation and put him on growth hormones and mothered and protected and babied him right up until the day he killed her. And later, when I get back home, I find a book in my library called *Freaks: Myths and Images of the Secret Self.* Written by a brilliant literary critic named Leslie Fiedler, it tries to answer the question of why we stare when we see a deformed person instead of simply looking away. His jumping-off place is a passage by Carson McCullers: "She was afraid of all the freaks, for it seemed to her that they had looked at her in a secret way and tried to connect their eyes to hers, as though to say: *we know you. We are you!*" Fiedler argues that this sense of secret complicity begins in early childhood, when we are all tiny creatures in a world of giants. Just as we reach the threshold of the big world, we see parts of our bodies change and swell and grow secret hair that is "more like animal fur" than normal decent hair and begin to sense, with a complicated mixture of resentment and pride, that we are wild things "more like our pets than our parents." That's why classic children's stories like *Alice in Wonderland* and *Gulliver's Travels* show the hero or heroine changing shape until all the wrong shapes are rejected and the right one found, a solution that brings relief from our own freakishness, a relief we call maturity. That's why we stare, because something in us *needs* to stare, because we still have our secret fur to remind us that parts of us are so disturbing to decent and normal society that they must be hidden from sight, because our sexual differences ensure that men and women be "forever defined as freakish in relation to the other"—because even though we may have grown up in a time when men were growing their hair long and women were leaving their legs unshaven and Frank Zappa said "freedom is wearing a dress" and French philosophers became so obsessed with difference that they spelled it with an "a" to make it more différant (all those small and now faintly ridiculous changes that nevertheless added up to a genuine change in our relationship to the unusual)—it's hard to keep that faith. Because (Fiedler argues) the sight of a deformed person is a kind of primal religious experience that threatens the division between weakness and health and beauty and ugliness and finally even the division between Self and Other. We stare because we are transfixed by "quasi-religious awe" at the sense of "the final forbidden mystery"—by the Other that might just be Us. I bought Fiedler's book in college, and I've been dragging it around for fifteen years, from state to state and apartment to apartment. Later I'll read another version of these ideas in a book called *Human Disability and the Service of God*, which argues that feeling giddy or dislocated or disturbed in the presence of disability is not only common but a twitch of humanity waking up, that we need the disabled to snap us out of the fascism of health and remind us of "the power of life and its brokenness." So I have to consider the possibility that it's no accident that I've dragged this tattered copy of *Freaks* around all these years and no accident that I've never finished reading it and no accident that I'm here at this convention. Because I do judge people by their looks. Because odd moles and stray hairs disturb me. Because the plain and fat have always had to work a little harder to get my attention. Because the sight of a beautiful woman stabs me like a spear and makes me feel ugly and ashamed.

And here I am, sitting in the lobby of the Marriott hotel with the dwarf-shock still hitting me.

. . .

As the afternoon stretches on, there's definitely a tension building. A little man passes through complaining about the lack of air-conditioning. "Maybe they figure we need less air," he snaps. Someone else is fuming at the hotel's photoelectric door, which won't open for him. The beam isn't set to read someone his size, so he can either jump around trying to set it off or wait for a tall person to come along.

Nearby, some teenage boys gripe about the short life. "Come on, guys, it's not that bad," says one flirty girl.

"Yeah," comes the answer, dripping with teenage irony. "It made me what I am today."

Outside the hotel doors, I come across a stocky dwarf pushing an angry thin dwarf against a wall. The stocky man is one of the convention organizers, Barry Phipps. He's telling the skinny guy he can't treat people like that, can't talk to people like that, and the skinny guy is saying "leave me alone" and trying to squirm free. But Barry jams his forearm under the guy's chin and tells him he's going to have to leave the convention and not come back—not today, not tomorrow, not for the whole week. A few minutes later, as the skinny guy skulks off across the parking lot, Barry tells me the trouble started when the other man got carded in the hotel bar. Sure, he's a grown man, thirty-two years old, and it can be a real drag when clueless tall people treat you like a kid just because you're short. But that doesn't give you a right to call someone a nigger.

. . .

Michael Gilden? Kay hasn't seen him, but she knows him well. She's thirty-three, with huge eyes and a sweet, hesitant manner, and happy to talk. She's never had a relationship with a little person before, except a pen pal when she was twenty-four who tried to convince her that they should be together just because they were both little. Which didn't sit too well. Then it was a long stretch of nothing. Until last year. She says this sitting in the comfy club chairs in the lobby, as the afternoon shadows start to fall across the high windows. Of course she feels like she missed something being alone all through her teens and twenties, but she was so overweight! Right now she weighs ninety-three pounds, which is just about right for someone four foot three. But picture her weighing 170 pounds. A lot of little people have that problem, eating "average" portions and not exercising because their joints hurt. When a friend died of a heart attack at thirty-three, Kay finally joined Weight Watchers. She came to that first convention a lot skinnier but still feeling so raw and tender, and that's when she met Michael Gilden. He seemed so sure of himself. And he was nice to her, explaining the anthropology of the LPA. She took a lot of comfort from him. Then she started hearing the rumors about his womanizing. Maybe it's because he's so good-looking, or because of the way he carries himself, so confident and bold, like he's six feet tall, but a lot of people seem to resent him. Only months later did she start meeting other people who said he was really a good guy and he was just misunderstood, so she felt bad for believing the rumors and called him, and he turned out to be a very supportive and loving friend—her best little person friend. He pumps her up for the conventions, which she needs because she has more confidence in the outside world than in the little world. "It's a shame, really, that people treat him the way they do. He really is a wonderful person. He's not as confident as he appears to be, you know. He's got the same insecurities as we all do."

It's funny, she says. Aside from Michael, all her best relationships are with tall people.

"You've had romances with tall people?" I ask.

"My first relationship ever was with a tall person last year," she says. "I was the one who broke it off."

. . .

Then Michael comes down to the lobby in a blue tank top, his Nike beret turned stylishly backward, looking as casual as a man can look. But inside he's a ball of tension. He went through a bitter divorce a few years ago and ever since he's been really vulnerable, and this girl he was telling me about, the one he met in a chatroom on America Online, has really got his hopes up. Her name is Meredith Hope Eaton. She's from Long Island, a judge's daughter, and she's funny and smart and really pretty too—he knows because they traded pictures through the mail. One day he even sang to her, and it was no accident the song was called "Open Arms" (by Journey, his favorite band). It took forever to convince her to come to Atlanta to meet him. Problem is, she's always dated tall guys. She's never even seen a little man before. And now she's about to see hundreds of them.

Michael winces. Michael scowls. Michael looks at all the dwarfs and talks, pouring out his anxiety in a flood of words. These conventions are an alternate reality, he says. The whole little *world* is an alternate reality. When somebody like Meredith walks into this place for the first time, plunged into this world where small is the norm instead of the exception, she's just naturally going to find it upsetting. *Very* upsetting. Because people adapt. They adapt to the tall world to the point where they almost forget they're small, and they don't want anything to remind them. And when they see other dwarfs for the first time, especially in large numbers like these, they just plain freak out. They see the big butts and big heads and little arms and little legs and it hits them like a truck: Do I look like that? A lot of times, they walk right out the door and never come back.

This is what Michael has been brooding about all morning, the fear that's branded that frown onto his forehead. "She's never experienced anything like this," he says. "She's gonna be terrified."

Then Meredith walks in. She's got black hair and black eyes and a perfect heart-shaped face. She's wearing a black DKNY T-shirt and a thick gold chain, sunglasses tilted up on her head, looking exactly like the classic Long Island born-and-bred Jewish-American princess she is . . . except her arms and legs are short and sausage-bulgy, and she's leaning on a silver hospital cane, and her hips jog to the side with each step she takes, giving her a slight waddle.

Then she stops. She's frozen in place, visibly terrified.

A tear comes to Michael's eye.

Then she turns around and walks right back out the door.

A minute later, she comes back. This time Michael hurries up to her. "You made it," he says.

"Barely."

But she won't go any farther into the lobby. She *can't.* And she doesn't want to walk back out again because that would be ridiculous. So she takes a seat right next to the door in a deep club chair. Her mother sits down right next to her. "This is just so weird," she says.

"How's the surgery?" Michael asks.

She points to her right leg. "It was this one."

"They look straight," he says.

"They should be, after all I went through."

Michael smiles with infinite kindness. "You'll do some dancing tonight," he says.

. . .

Barry Phipps says they're all true, the things I'm hearing about the LPA conventions and romance. He met his own wife at a convention, seven years ago this week. And once that was out of the way, they had to face the next big dwarf relationship issue, deciding whether to have a child or to adopt. Dwarfs who mate with other dwarfs have a 50 percent chance of giving birth to a dwarf, a 25 percent chance of giving birth to an average child, and a 25 percent chance of giving birth to a "double-dominant" child so deformed it dies at birth or shortly after. They also know that some average-size couples don't want their dwarf babies and put them up for adoption, and a dwarf just naturally feels for kids like that "because they know what it's going to be like—like the medical problems and how their clothes fit and when they're teenagers and all their friends are getting some and they ain't gonna get any." So they flew to Paraguay and met a pair of unwanted dwarf twins, but that didn't quite work out—too many problems and too much red tape. Finally they decided to go for it. Now they have a three-year-old girl who's already eye to eye with her mother.

As Barry tells me this story, he sees a small brunette woman and calls her over. Her name is Nancy Rockwood, and she runs the LPA's adoption office. She tells me that adoption is as touchy a subject in the little world as religion or politics. "There are two schools of thought—one is that you shouldn't pass on 'bad' genes, and the other is, 'I'm going to have mine no matter what.' " As to availability, Americans have gotten pretty good about keeping their dwarf babies. On average, only one dwarf child gets put up for adoption each year. So nowadays most of the unwanted ones come from overseas. "They don't want them if they have a cleft lip," she says.

. . .

Meredith was about six when she first noticed that her hands and arms weren't growing right. She cried and cried. Middle school was when the teasing began, the worst time of her life. One kid used to harass her every day. But she stuck it out and blossomed and in high school she met a guy, a six-foot wrestler who adored her and protected her and treated her like a fragile china doll and never made her feel bad about her height. She dumped him for a guy who was six foot four. Maybe it was stupid, but she was at that age when she was exploring the world and testing her limits. Only this time she learned that four extra inches wasn't the solution to everything. When they were alone, the new guy loved her, but when they went out in public, "he'd look around, and when people would stare at me, he'd laugh nervously and go, 'Isn't she beautiful?' to try to mock them from staring. He never actually said it, but I knew that inside it really bothered him. We went out for three and a half years, and we must have gone out to dinner twice."

There's something instantly comforting about Meredith, a warmth that almost shimmers. She spent her summer vacation working with severely disabled children, and she's thinking of making it her career. She's working toward a doctorate in psychology. It's just that she's having so much trouble with this height thing. Before she came to the convention, she had nightmares where she'd be walking her dog—a miniature poodle—except in the dream he was even smaller and she lost him and searched and searched and couldn't find him anywhere. And she'd see very tall people, who for some reason were always flying. And bugs too, lots of itty bitty ugly bugs.

Twenty minutes have gone by, and she's still sitting by the door. "I'm the *same* as everybody else," she says, sounding surprised and a little forlorn. "I'd feel more comfortable if this whole lobby was all people average-size."

"You talked on the phone," Michael says. "You were totally comfortable with that—"

"That was different."

Another half hour passes, and she's still sitting there, still visibly uncomfortable. "I want to say I don't look like these pe— like everyone else. It's something that takes getting used to."

Meanwhile, Michael's mooning at Meredith like a lover in a silent movie. He's mooning and gazing and mouthing sweet nothings across the table.

"What?" Meredith asks.

"It's just everything," he says. "The face, the eyes, the twinkle, the smile—everything is confirmed."

"Stop," Meredith says. "You're embarrassing me."

. . .

Later, Michael is in a fever. "It's not like, 'Well she's kind of cute.' I *can't stop looking at her.* I don't want to play volleyball. I don't want to see Atlanta. I don't want to do anything. I can't believe we're going to be in the same room breathing the same air! And she's—she's had three-, four-year relationships with average-size guys. I don't even know if she can *look* at me."

And yes, he knows that would be poetic justice, because back when he was a teenager coming around for the first time, he'd look at the little girls and turn away. He was cool with the guys, but the women just looked wrong. They looked *deformed.* Oh man, he'd think, that doesn't look right. Maybe it was because he grew up in an average-size family and had a brother who was six two. It's what he was used to. But eventually he did start dating little women and got used to that and even married one, and just when he was all set to start a family and live happily ever after in the little world, his marriage fell apart. For Michael, it was the ultimate blow. "And I tried to recover from it," he says, "and my recovery was I needed *her*—I never knew who she was, but I needed her to show up—and for the last three years she has not shown up. And now she's here. Without a doubt, without a *doubt*, this is a girl I want to develop something with, I *know* it, and I'm wrestling with how to behave, because there's a part of me that wants to right now sit and have a drink and have dinner with her *alone*, and I fear that because of her feelings, and because of her mom being here also that the likelihood of me doing that this week won't happen—by the time she's comfortable she's going to be leaving, and I go three thousand miles away. . . ."

I try to calm him down. I tell him they look perfect together, and they do. I tell him she seems attracted to him, and she does. But it does no good. The thing is, Meredith isn't the first person he's met this way. Last year there was a girl named Carrie. He met her in the same dwarfism chatroom and things were shared, stuff like her having an abortion and his thoughts about adoption—they even went grocery shopping on-line, a practice run for real life. But she wouldn't take the next step, wouldn't even talk on the phone, and when he tried to compromise by sending her voice files so she could hear him speaking, she wouldn't send any back. She never even sent a picture. And though she promised to meet him at the convention, she never showed. And now he's thirty-four years old, and he wants kids and a family just like anybody else—just like any tall person. He should *have* kids by now. And if it doesn't work out with Meredith . . .

"The pivotal thing that will happen tonight, and I think this is extraordinarily pivotal, is when the dancing starts," he says. "She's not going to fast dance because of her legs, but I know either Dave"—Michael already has competition, a kindly, balding dwarf who also met Meredith on the Internet—"or me or someone will approach her to dance slow, and she will at first probably say no—but the first person she says yes to *has to be me!*"

. . .

That night the dwarfs gather in the ballroom, many of them glammed out with teased hair, ball gowns and serious cleavage. The guys come in doing the dwarf strut, chest out and arms back. Soon the dance floor is filled with little people getting down. One little guy kneels on a kind of tricycle, leaning forward against a tilted pad and shaking his butt in time to the music. There's Kyle dancing in a white baseball cap turned backward, weaving in funky loops. And there's a tall girl dancing too, willowy and slim, rising above the boogying dwarfs like some kind of goddess. After each dance she leans down to embrace the small men and kiss them. I begin to think of her as an Emissary of Tallness, carrying the good wishes of the large to the small. Even though she's just somebody's teenage daughter, her long arms and willowy legs look so glamorous.

A few minutes later Michael shows up dressed to dance, dapper in khakis and a black shirt, looking like a classic L.A. club dude. But inside he's still a ball of anxiety. In the lobby earlier, he tells me, he asked Meredith straight out: "You don't see me the way you see your other boyfriends, right?"

Meredith dodged the question.

Now Michael's radar picks up movement across the room. "Here she comes."

Meredith is leaning on her cane, working her way through the crowd. She's wearing a silvery gray shirt split from one button to expose her belly, which is brown as a walnut and supermodel flat. A silver clip gleams in her black hair.

Michael leads her to an empty table, whispering into Meredith's ear. Moments later, a dwarf named Kris joins them and introduces his new girlfriend, Lila, a pretty, poised eighteen-year-old with puffy blond bangs. "Those are beautiful shoes," says Lila.

"They're painful," Meredith answers. "Shoes are always painful—beautiful but painful."

Plus the humidity is hurting her leg, so she doesn't feel like dancing just yet. Trying to be nonchalant, Michael forces himself to dance with other women.

Taking advantage of the dark, I study Meredith and decide she looks a little bit like the queen in *Alice in Wonderland*—not her face but her body, which is squat and a bit wide. Then she notices me looking at her and gives me one of her dazzling smiles. "I feel better than I did this afternoon," she says. "I really, *really* like talking to people at eye level."

Then the DJ plays a slow song, and the tall girl slips down to her knees on the crowded dance floor, resting her head on a little man's shoulder. Again, I feel the warmth that comes when the rules of life are suspended and reordered in a kinder way. When the music kicks into high again and the tall girl gets up to boogie, it's giddifying, like somebody stuck those damn rules in a blender and hit frappé. Then the tall girl bends over and kisses the dwarf and it's all so swee— Whoa! That's no Emissary of Tallness kiss! There's some serious tongue involvement going on there! This willowy tall girl has a foot and a half on this little guy and she's bending over to stick her tongue in his mouth!

This is getting confusing.

Two

THE OLDEST KNOWN DWARF LIVED IN ITALY about ten thousand years ago. When his bones were discovered, anthropologists took it as a sign of tolerance that he survived to late adolescence, a challenge to the general assumption that ancient people feared deformity as a curse sent by the gods. But that was in the 1960s, when racial and sexual differences sparked such vast political upheavals and academics were on the hunt for new paradigms. The broader evidence suggests that tolerance went hand in hand with dread, that our ancestors were both fascinated and frightened by what scholars call "liminal" or "threshold" creatures. On one side, it sometimes seems that the ancients saw in every blind man a potential prophet and in every beautiful woman a nymph. The early Greeks even allowed lepers into some of their temples, the priests having ruled that purity of mind was the important thing. The Roman poet Ovid wrote a long romantic book about "bodies changed to different forms" and "the gods that changed them." But the same ancients dreamed up dogs with three heads and women with snakes for hair and lions with human faces, nightmarish visions that expressed our dread and posed deadly riddles about the nature of man—it's no accident that the Sphinx asked that famous riddle about what had four legs at morning, two legs at noon and three legs at evening, and also no accident that the answer—man—required an insight into human mutability. Strange bodies posed classic questions of fate and destiny and chance. That's why all early religious texts had complex rules and prohibitions for dealing with any person who is diseased or disfigured or deformed. Should we accept what is given? Should we offer up a sacrifice? Or should we just kill the hideous thing and hope it was a mistake? No surprise then that the Egyptian god of childbirth took the form of a dwarf, or that dwarfs appeared in sculptures and frescoes and bas-reliefs from ancient Greece and Pompeii to pre–Columbian America. There's even a dwarf Egyptian mummy, Knoumhotpou, Keeper of the Royal Wardrobe. And a man "little of stature" in the Bible, climbing a sycamore tree to peer at Jesus.

Jesus spent the night at that little man's house, his answer to the riddle of difference. The Romans had another answer. For them, dwarfs served as court jesters or glorified pets, fighting in the arenas as comic gladiators, a perfect expression of the Roman taste for cruel and fatalistic jokes about destiny. Eventually their appeal was so strong that some Romans began creating them artificially by starving children or resorting to exotic "dwarfing recipes" made from a brew of dormice, bats and moles. In China, dwarf jesters were in such demand that artificial dwarfs were made (so the horrible story goes) by raising normal children in jars. For the next two thousand years, the dark joke kept working. Catherine de' Medici kept six dwarfs as "pets." Charles IX of France got three as a gift from the Emperor of Germany. King Sigismund Augustus of Poland had nine. An Italian noblewoman named Isabelle d'Este collected so many she built a separate wing on her palace just to house them. Peter the Great had more than seventy. Dwarfs were deployed with humor or sentimentality, as when thirty-nine dwarf waiters scrambled to serve guests at a sixteenth-century banquet, or when Peter the Great celebrated a royal wedding by marrying off two of his pet dwarfs (the wedding party included more than seventy little guests). In the court of King Charles I, a dwarf named Jeffrey Hudson was presented to the queen "baked" under the crust of a cold pie. Artists like Velazquez, Goya, Raphael, Veronese, Van Dyke and Brueghel treated them much the same way, painting them with almost compulsive interest, but usually posed with monkeys or dogs.

All along, certain dwarfs distinguished themselves, stubbornly rising beyond the celebrity of size. In ancient Greece, the fabulist Aesop is believed to have been a dwarf. The philosopher Alypius was said to be only seventeen and a half inches tall, yet wrote one of the first studies of musical scales and notation. In the sixth century, a dwarf named Gregory of Tours wrote an important early history, *Historia Francorum.* In the late Middle Ages, a dwarf named Bertholde became prime minister to the King of Lombardy. In twelfth-century Egypt, the emperor Paladin had a four-foot minister named Characus. In the fourteenth century, a dwarf king named Charles III ruled Naples and Sicily. King Ladislas I ruled Poland and was so brave he became known as the "warrior midget king." In the fifteenth century, John de Extrix was a distinguished scholar and aide to the Duke of Parma. A century later, Cardinal Richelieu appointed a dwarf named Godeau as Archbishop of Grasse. From 1830 to 1850, the position of messenger of the British House of Parliament was held by a three-foot man named George Trout. During the French Revolution, a dwarf spy named Richebourg dressed as a baby to smuggle messages in and out of France.

Even some of the court pets distinguished themselves beyond their size. A talented English dwarf named Richard Gibson painted miniature portraits so skillfully they kept him in favor through three revolutions; his sitters included Charles I, Oliver Crumble, and King James II. While serving as the king's courier to Scotland, Jeffrey Hudson was taken prisoner by pirates (and ransomed) three times. When a nobleman mocked his size, Hudson challenged him to a duel. When the man mocked his challenge, Hudson whipped out a little gun and shot him dead. Late in the eighteenth century, a Hungarian dwarf named Josef Boruwalski published a best-seller based on his love letters to a tall woman named Isalini, later his wife and even later his ex-wife (and how everyone laughed at the droll story of the day she finally perched him on the mantelpiece and walked out).

The lives of more ordinary dwarfs are mostly lost to history, but the Bible tells us the little man who peered at Jesus from the sycamore was "chief among the publicans." In ancient Greece, dwarfs served as medical assistants, a role in tune with their liminal magic. A dozen odd centuries later, the sixteenth-century astronomer Tycho Brahe used a dwarf named Zep as a technical assistant and occult medium. Records from the nineteenth century show that at least three dwarf soldiers fought in the American Civil War, one just three and a half feet tall.

Then came the modern innovation that brought the mystic dwarf-jester tradition to the masses. Shortly after the first modern circus began with a horse act in London in 1768, dwarfs took their place as one of the main attractions. Before the century was out, the "Polander Dwarf" came to the United States and rode a horse through a flaming hoop to fame. By the 1840s, dwarfs had become such a popular attraction that P. T. Barnum was able to turn a Connecticut carpenter's son named Charles Stratton into one of that century's biggest stars. Performing under the name General Tom Thumb, Stratton updated the old dwarf-gladiator gag by dressing up like Samson, Hercules, and Napoleon and reenacting their most famous adventures. Barnum launched their first tour of Europe by wrangling an audience with Queen Victoria, who was so charmed by Stratton—his mock sword fight with her poodle became a tabloid sensation—that she had him back twice, introducing him to the Prince of Wales and sitting through an impertinent performance of *Yankee Doodle.* She even kissed him, starting a fad that brought to the surface an erotic fascination with dwarfs that persists to this day in the surprisingly popular dwarf porn genre. A year later Stratton left Europe with boasting rights to the lips of the queens of France, Belgium and Spain, as well as "nearly two millions of Ladies."

But the circus and the Victorian curiosity it expressed marked a turning point. As science and show business began to evaporate the lingering sense of dwarf mystery, the Little World began to stir. In 1898, a group of Barnum & Bailey performers who held a protest meeting in London calling for better treatment and a substitute name for "circus freak." Although this was probably another P. T. Barnum publicity stunt (the minutes were recorded by the Armless Wonder, using her feet), it resonated with the public, who submitted nearly three hundred substitute names. The protesters finally settled on a solution offered by the Bishop of Winchester—"prodigies."

It didn't catch on, but other trends were rapidly changing the way people thought about "liminal" humans. Like dwarfs, people with various disabilities—the blind, the lame, the dumb, and the mad, even people with exotic diseases like tuberculosis—had been alternately shunned and draped in mystery. In colonial America, people with disabilities were considered unfit for the rigors of the New World and summarily deported. But after the Revolutionary War, the new government recognized its debt to soldiers crippled by battle and began paying them pensions. In time, this gave birth to the Veterans Administration hospital system and the Public Health Service, both to become major players in the disability movement. A new idea was stirring in Europe too, that instead of shutting away the blind and the deaf in almshouses and attics, instead of assuming they were retarded or unteachable, to try to integrate them into society. The idea made its way to America and flourished; in 1812, the first school for the blind was built in Maryland. Five years later, a Connecticut reformer started the first school for the deaf. In the 1840s, a schoolmistress named Dorothea Dix began an almshouse reform movement that shamed America with tales of chains and beatings. The Civil War had a huge impact too, producing such a mass of crippled soldiers that one year the state of Mississippi spent 20 percent of its revenues on prosthetic arms and legs. By the time World War I rolled around, new medical techniques helped even more young men survive terrible wounds, draping disability in an aura of heroism and patriotism. For the first time, Congress began funding physical rehabilitation programs for veterans. By the time Franklin Roosevelt pushed through the Social Security Act in 1935, there was enough popular support that he was able to include permanent financial assistance to all disabled Americans.

All this fueled a new spirit of activism. During the Great Depression, a group called the League for the Physically Handicapped occupied the WPA offices in Washington to protest job discrimination. In 1935, a pair of visionary ex-boozers named Bill Wilson and Dr. Bob started Alcoholics Anonymous and launched the support group movement. In 1946, the National Multiple Sclerosis Society set up shop, followed by the Cerebral Palsy Association in 1948 and the Muscular Dystrophy Association in 1950—and by new groups for the blind and the deaf and polio survivors and a vast number of other ailments.

And finally, one day in 1957, a dwarf actor named Billy Barty appeared on a TV show called *This Is Your Life* and used the national forum to suggest dwarf solidarity. At the first meeting in Reno, twenty-one people from nine states gathered under a banner that said WELCOME MIDGETS! At the next meeting, more than a hundred dwarfs showed up.

The Little World had begun.

. . .

Sunday morning, Evelyn and Jocelyn make their way to the hotel restaurant. At the buffet table, Evelyn notices another platform and feels a spark of pleasure at the sight of the tall people awkwardly navigating it, standing with one foot up and one foot on the floor. Good, she thinks. Jocelyn has always been inconvenienced. Let the tall people suffer for a change.

The funny thing is, now Jocelyn feels out of place. "I'm used to being in a tall world," she tells her mother. "This is weird."

Crossing back to the elevators, they find the lobby swarming with little people. Now it's Evelyn's turn to feel weird—full as the room is, she can see across it with ease, peering over the heads of the dwarfs like an airline passenger looking down on clouds. There's just one other tall head poking up above the crowd—a smiling man who is bending down and hugging one dwarf after another. She asks one of the dwarfs who he is and he tells her that it's Dr. Steven E. Kopits, the famous dwarf specialist, information that sets Evelyn's heart pounding. Dr. Kopits is one of the main reasons she came here. Last year, when Jocelyn's headaches started getting so bad, they called him from Australia. He wanted to see her then, but they didn't have the nerve to make the trip. When she found out he'd be at the convention, giving free clinics, it tipped the decision to come, but she still held back a bit by not making an appointment in advance. Somehow that made it easier to face—she could concentrate on getting herself there and worry about the rest later. And now that time has come. She hurries across the lobby and waits for him to finish his latest hug-and-hello and when he turns his gentle blinking eyes on her, she rushes out the abridged version of their story. He immediately agrees to see them later that afternoon.

That's when I come up to introduce myself. They're on their way to church with Barry Phipps and his wife Bunny, and Barry suggests I tag along. No problem, Evelyn says. She'd love it if more people knew what families like theirs had to go through. Jocelyn nods in agreement. So we all clamber into a van and drive to the biggest Pentecostal church in Atlanta where the choir is supposed to be huge and amazing. On the way, Evelyn tells the story of discovering the LPA website that fateful day in April and their decision to come to America and how impressed they are with all the arrangements the hotel made for the little people. "I *love* it that tall people are inconvenienced," she says.

At the church, we get stares. As an attendant leads us down to the very front row, Barry and Bunny and Jocelyn all look straight ahead with barely a glance to either side, as aloof as movie stars. Evelyn too. When the ushers leave, Bunny turns to face her pew. She's three feet eleven inches, so the seat comes up to the middle of her chest. Like a child, she has to use her knee to scramble up.

For the next hour, a tag team of big-throated gospel singers warms up the audience, backed up by a full band and a choir of sixty. Then the preacher takes the stage and starts telling us that "the enemy" has crippled us in body and spirit and what we need to do is start thinking about our bones. "Without bones we'd just be slithering creatures," he shouts. "God has put into our bones genetically the structure of our lives! Your bones, your *bones* have a testimony to tell concerning your destiny."

Sitting between Jocelyn and Bunny, I squirm. Doesn't he see them? How can he say these things with them sitting there? But he rolls on heedlessly. Did we know that there are twenty-six bones in the foot? Did we know that proper alignment of the bones relieves stress and pain? Did we know that forensic anthropologists can tell from bones whether a caveman smoked a pipe or labored with his hands, whether he stood erect or hunched? "A bone is strong and dependable, and you never really realize it's there until something goes wrong," he says. "Only when it is fractured do you then realize that its strength and dependability has served you well." His point is that God and the church serve us as the bones serve the body, and "the reason we're experiencing the recent health crisis in America and the reason why we're experiencing a skyrocketing divorce rate nowadays is because we've rejected divine order—because our *bones* are out of place in our personal lives."

Using the body as a metaphor is as natural as breathing, of course. We all do it. But as the preacher gases on endlessly, stretching his metaphor way past the breaking point, popping ankles and iliums into every possible socket, it begins to seem so cruel. In her book *Illness as Metaphor*, Susan Sontag wrote about how dangerous this kind of thinking can be, that the old idea that disease was divine punishment comes back as the idea that cancer is caused by repressed anger, the "revolt of the organs," with the result that we are so ashamed or fatalistic about illness that we avoid getting treatment. Worse, our need to make meaning out of illness pushes us to weave "punitive and sentimental fantasies" around it. It's no accident that one of the oldest stories about the origins of dwarfs has God making them after Adam and Eve, from the leftover clay, with the inevitable implication that they are not quite fully human. And despite Jesus' sleepover with the little man, the Bible almost invariably views illness in the old judgment-of-God mode, treating leprosy and blindness and other physical defects as punishments we have brought upon ourselves. In the Book of Mark, Jesus heals a paralyzed man with the words, "Your sins are forgiven." In the Book of John, Jesus heals a weakened man with the words, "Do not sin any longer, so that nothing worse happens to you." In the Book of Leviticus, the same harsh text that orders homosexuals and adulterers put to death, priests with physical defects are forbidden even to approach the altar of God. No wonder that popular culture echoes these stigmas, and books and movies from the Brothers Grimm to Disney show sweet fairies proportionate and nasty gnomes bent and "dwarfish." From there it's not such a long step to Hitler's camps, where dwarfs took their place alongside Gypsies and Jews, slaughtered for a fantasy of human perfection.

Then the preacher reaches his climax and tells us that there is another truth hidden beneath the skin—that we are all connected, like bones in the body of the one true church. "You may not think we are connected, but we are," he says. "So turn to your neighbor and say, 'Hello lovely bone.' "

I turn to Bunny. "Hello lovely bone."

She scowls. "You don't know what you're getting into," she says.

. . .

Evelyn and Jocelyn's room looks like an airport bomb exploded. Clothes cover every surface, shoes fill every corner, suitcases bulge open on the crowded floor. Jocelyn is out of her wheelchair, moving around the room in an oversized T-shirt. To help pass the time while she waits for her appointment with Dr. Kopits, Jocelyn tells me about growing up in Australia, the stares and the giggles. "You've got to tell everybody you're not five and you're not in kindergarten, that you've actually got a brain in your head—there's actually a person in here. I can talk."

"Then she has to deal with the do-gooders," Evelyn says, "the ones who want to wrap her up—"

"In cotton wool."

"And she's got to follow what they want to do, and that's really hard to break free of, because they couldn't understand that they were just completely squashing her whole life, her expressions, her freedom."

Evelyn is so forceful and passionate about this, it's almost as if it happened to her instead of to her daughter. But as she tells me her story, it becomes more and more clear in how very many ways mother and daughter overlap. Born forty-three years ago in a small town outside Sydney, she was the youngest child in a family with three older brothers. As she describes them, her parents were dour and strict immigrants who followed the rules of the Dutch Reform church to the letter. Dating was discouraged along with fashion, television and other frivolities. On Sundays, they didn't even use the telephone. After high school she studied nursing and fell in love with David, a young engineer who was not Dutch Reform. Her parents objected, and Evelyn pushed their limits even more, living "in sin" with David for a brief period. But they got married and quickly started a family, producing a beautiful and healthy daughter they named Alecia. Then Evelyn got pregnant again. Six months in, they knew something was wrong. Evelyn was too big, huge, as if she were carrying a baby giant. The doctor thought the baby might have a hole in her esophagus that caused her to take on excessive amounts of fluid, so Evelyn spent the last two months of pregnancy in the hospital. The birth was terrible, long and painful, and at the end Jocelyn was born blue, almost strangled by the umbilical cord.

For the first twenty-four hours they didn't know if she'd survive. The next day the doctors said they thought Jocelyn might have hydrocephalus because her head was so big. Evelyn noticed a tape measure by her crib.

After a week, the hospital sent Evelyn home and kept Jocelyn for supervision. A week after that, the doctor called Evelyn to his office, where he put Jocelyn in a baby basket and measured her from fingertip to fingertip and from head to toe. No matter what your age, that should always be the same length.

"Really?" I say, stretching out my arms. The arm span is that long?

"Every time I tell that, people do that exact thing," Evelyn says. She mentions Leonardo da Vinci's famous drawing of a man in a circle—that's the principle he was illustrating.

Jocelyn's arms were each about four centimeters too short. If da Vinci had put her inside his circle, her fingers wouldn't have touched the edge.

The doctor opened his medical textbook and showed her the diagnosis.

Evelyn asked him, "Is it because I didn't drink enough milk?"

No, nothing like that, the doctor said. He explained about genes and how these things happen and that nobody is to blame.

That night driving home, Evelyn's tears came so hard she thought it was raining and turned on the windshield wipers.

When she told David, he seemed stunned and didn't say much. He tended to be restrained in his emotions, so she let it go. But the next day, he still seemed so strangely emotionless. "I remember looking at him and thinking, When are you going to do something? When are you going to get angry? When are you going to cry?"

Later that day, she dug out some of her old nursing textbooks and found an entry on achondroplasia and learned that the word comes from the Greek for "without cartilage." The distinguishing characteristics are short limbs, bowed legs, an average-size trunk, swayback, a flattened nose, a disproportionately large head and short-fingered hands with a wide separation between the third and fourth fingers (known as "trident" hands). As the doctor said, it was a "spontaneous genetic mutation"—a fluke.

Then she found this passage: "Adults with achondroplasia often earn their livings as clowns in the circus." Tears blurred the words. Is that our future? she wondered. Is that what Jocelyn has to look forward to? She closed the book and hid it under the bed and dreamed all night of her baby daughter wearing a big red nose.

That was the beginning of a terrible year. She called her parents, and her parents said, "When are you coming back to church?" David's parents said, "We need a second opinion." That was when she left the room for a few minutes and came back to find David crying. "I thought, Good—not because he was crying, but I knew there was somebody *there*." Evelyn's first impulse was to have another child, have another one right away, and she got pregnant quickly. Even so, she and David barely talked. He worked long hours at the power company and she spent all her time at the hospital, communicated by messages on the kitchen corkboard, until finally they landed in the office of a marriage counselor. And Jocelyn did nothing but scream all year long because she couldn't lift her heavy head. Evelyn would feed her, walk her, do anything to get her quiet, and after a while it all began to seem like a dark sleepwalking dream. "I remember folding nappies over and over and over, and they were still too big," she tells me. "I remember trying to pull these legs—there was just no leg to come through, the nappies were so big, and I just kept yanking these legs that just weren't *there*."

But they got through it. And at the end of the year Evelyn gave birth to a boy, Nicholas. And eventually things began to seem fairly normal. And normal was the way Evelyn wanted it—they went once to the Australian Little People organization, but she didn't like it. Better just to raise their child in their family, in their world.

It worked until Jocelyn got a bit older. When she was about three and a half, she developed a habit of lingering over meals, and one day she sat and sat at the table, asking for more even though she could barely stuff it down. Finally Evelyn asked her why. And she said, "Because if I eat more, I'll grow up tall like my big sister." So Evelyn took a deep breath and said, "No darling, that won't happen, because you're a dwarf." And they sat at the table for a long time and explained what that was.

Soon the physical difference was apparent to everyone, and Jocelyn's schoolmates began wrapping her in that cotton wool. I ask Jocelyn to describe their part and she says she had to keep fighting to be treated normally. But she doesn't sound angry about it, just a bit annoyed at how silly the world can be. Nor was she aware of being particularly angry when it happened. The first moment of real anger came when she was about ten and saw another dwarf for the first time: I don't look like that, she thought.

It took a few days to adjust.

Going into the wheelchair was the next big problem. They had to sell their house and build a new place with bigger bathrooms and no stairs. Then Jocelyn started having trouble with her classmates at school and started to become more and more of a loner. When she started using an electrical wheelchair, the process was complete. "People really started to treat me like I was brain-dead then," she says.

With all the turmoil, Jocelyn's doctors suggested that the whole family go to a therapist, which sparked their first serious rebellion against medical authority. Evelyn was the first to resist, bristling at the way the therapist kept harping on her childhood troubles with her parents. "I couldn't see the relevance," she tells me. "All of my life, all of our lives since we had Jocelyn, if we can deal with something, it makes me feel better. If we can buy some clothes that fit her, it's great. If we find a pair of shoes that fit her, I'm really happy. We're moving forward instead of backward. And he wanted us to go backward."

When the therapist tried to get Jocelyn to talk about her troubles at school, she resisted too. Echoing her mother, she explains: "Mum and I, we both put our problems away," she says. "We never deal with them, we just keep going. He wanted me to keep going back, and I was, 'No, I'm getting ready for tomorrow.' "

Then came the night Evelyn and Jocelyn were sitting together alone and Jocelyn started saying, "I don't understand why I'm a dwarf—it isn't fair." For Evelyn, it was a scary thing. She'd never seen Jocelyn act that way before and didn't know what to say. She couldn't just tell her to buck up and everything would be fine, because that would be silly. Things weren't going to be fine. So she decided they couldn't afford to muck around in the swamp of feelings and told the therapist to take a leap. Before long, Jocelyn shrugged off her schoolfriends too. "I've had a lot of experience with life that not many of them would have in their entire lifetime, so you can't really relate to them and you can't talk to them about your problems," she explains. "They talk about their little problems like at home, or boyfriends or homework. That's nothing to me."

Then the headaches started. They got so bad, she would pass out at her desk at school. Finally she lost control of her bladder, and the doctors ordered an emergency spinal decompression. That was another horrible nightmare and months in the hospital and *still* didn't solve the problem. When she got home, all she could do was lie in the dark and sweat with pain. Watching that was the hardest thing Evelyn had ever done. Especially since she'd sworn she wasn't going to give her daughter back to those butchers at the hospital ever again.

That was when she rang Dr. Kopits. She'd found his name on the back of a dwarfism information pamphlet the hospital had given them. Kopits told her it sounded like Jocelyn might need surgery again and suggested some tests, so she gave in and went back to the hospital—and the Australian doctors refused to do them! They said there was a lot of risk involved and they wanted to see how much worse Jocelyn would get first. How much *worse* she would get! Just thinking about it makes Evelyn furious all over again. "Their egos were attacked!"

That was the first time Evelyn made plans to come to America. She even got passports. But in the meantime, they'd taken Jocelyn to a chiropractor, and that seemed to give some relief. Then they took her to a naturopath healer who moved her hands over Jocelyn's head and neck and actually made the pain go away. And Evelyn really hated the thought of going back to a hospital, any hospital. Before they noticed it, a year slid by.

Then the headaches came back. They didn't know what to do. And one day in April, Evelyn brought home a computer. There they were, sitting at the kitchen table in Cambelltown, Australia, not quite the bush but on the fringes of it, a world of tin roofs and outback bullnose porches, and within minutes Evelyn figured out how to boot up Netscape and call up a search engine. She asked Jocelyn, "What should we look for? Should we try dwarfism?"

From her wheelchair, Jocelyn nodded yes.

And voilà, just like that, they arrived at the website for the Little People of America. There was so much information! Lists of doctors and specialists and books and articles and all kinds of advice and links to dwarf web pages and dwarf databases and a dwarf "listserv" message board and even dwarf chatrooms. It was almost miraculous—with one click, sixteen years of isolation was over.

"So what's it like, being here at the convention?"

"It's weird," Jocelyn answers. "I see them in the hallway, hundreds of me."

Evelyn looks at her curiously. "When you were younger, you used to say, 'I want to marry a dwarf man.' Do you still want that?"

Jocelyn shrugs. "It doesn't matter. My brother's tall."

Jocelyn is so unflappable, so unnaturally serene, I ask her how she's come to deal with it all: how has she resolved questions about fairness and beauty? Does she ever ask, "Why am I like this?"

"Well, I'm a Christian," she answers. "I believe God put me this way. Beauty? I don't know. I don't think there *is* any—just people's perceptions is all I can say. There's beautiful people from my perception may be ugly people from your perception."

This is when it strikes me that there's such a thing as being too well adjusted. She's only sixteen, for heaven's sake.

But this is definitely not the time to press the point. We talked away most of the day already, and she has other things on her mind. In just two more hours, she's going to see Dr. Kopits and find out what's causing the headaches.

. . .

Chase the beautiful nymph into the forest and you will get lost. Reach for the sprite in the pool and you will drown. Glance just once at the charmer with snakes for hair and you will turn to stone. These are primal fears, a hint of why we flinch from beauty and why philosophers from every school and era tried so hard to dissect it. In Ancient Greece, they were sure it had something to do with proportion. For Aristotle it required "order and symmetry and definiteness," and for Plato it was parts of the right size fitting together in a seamless whole. But Plato took it a step further and said the seamless whole gave us a glimpse of spiritual perfection, an argument extended endlessly by poets and also by early Christian theologians who insisted that human beauty reflected our creation in God's image. It was a powerful idea, speaking to our deepest needs for justice and for coherence between the visible and the invisible. By the Renaissance, the obsession with correct proportion had become so fanatic that artists and philosophers developed insanely elaborate laws of human aesthetics—the ear and the nose had to be at equal height, the distance between the eyes the same as the nose, the face divisible into equal thirds. The Leonardo da Vinci drawing Evelyn mentioned was drawn for a Renaissance best-seller called *Of Divine Proportions*, which argued that the human form was not only God's self-portrait but also his design model for the "innermost secrets of nature," an idea later turned to poetry in Keats's famous line about beauty being truth and truth, beauty. But this powerful and romantic idea had a dark side: if men were made in God's image, wasn't ugliness a sign of man's fall from grace? Wasn't that what really turned you to stone, the snakes and not the beautiful face? By Elizabethan times, ugliness was so often associated with evil that a distinguished man like Francis Bacon could argue without inhibition that deformed people didn't have real human emotions. There was also the faction of Christianity that considered beauty a mere vanity at best, at worst a lure of the devil, something beneath the consideration of spiritual people. Gradually, a bad conscience began to infect beauty and sparked screwy arguments like G. W. Hegel's claim that Greek statues are so beautiful because they emphasize "spiritual" qualities like the forehead. With modernity, social reformers like Count Leo Tolstoy began lamenting "the delusion that beauty is goodness," and science began to blast away at the old mysticisms; Sigmund Freud dismissed the love of beauty as a form of narcissism caused by sexual repression and thought it pertinent that the main sexual organs are "hardly ever judged to be beautiful." Then Adolf Hitler came along and reached back to the old romanticism, taking the worship of beauty to demonic extremes. The world recoiled in horror, and the revolution of difference was launched. By the 1960s, feminism and civil rights and gay rights and the rock 'n' roll rebellion set to work "subverting the hegemony" of traditional notions of beauty with torn jeans and long hair and hairy legs, until we reached the point where a beautiful woman like Naomi Wolf could argue (in *The Beauty Myth*) that beauty was merely a "currency system" designed by patriarchal capitalism to keep male dominance intact and make us all good little slaves of desire. Competition for the scare resource of male approval was driving American women to buy nearly fifteen hundred tubes of lipstick every hour and put potentially dangerous saltwater and silicone bags into their breasts more than 120,000 times a year. Supermodel worship was driving a wave of anorexia, and one nationwide poll found that 10 percent of American couples said they would abort a fetus if they learned it had a propensity to obesity. So isn't Jocelyn exactly right and wise to say that "beautiful people from my perception may be ugly people from your perception"? Couldn't you even say that this is the best part of America, the key to its receptive democratic spirit?

. . .

Down in the lobby, Michael tells me that Meredith finally did dance with him—to "Open Arms." And after the dance he got his first kiss. Meredith said, "This feels so right." And she wasn't just talking emotion, although there was definitely "kissing compatibility," she was talking physical reality, because sometimes life is about accepting difference and sometimes it's about celebrating similarity, and damnit they lined *up*, knees touching and arms reaching just far enough—not big octopus arms wrapping her up like an infant. They're *proportionate*. To each other. And Michael said, "After experiencing this with me, don't you want to swear off tall guys? Don't you want to say, 'I'm not dating another tall guy *ever*?' "

She didn't answer, but she didn't let go of him either. Now the big problem is time. Meredith's return flight leaves Wednesday morning.

. . .

First they take measurements, leading Jocelyn over to the cloth ruler taped against the wall. She's 124 centimeters high—just over four feet.

"When did you arrive?" Dr. Kopits asks.

"Wednesday after a twenty-one-hour flight," Jocelyn answers.

"Twenty-one hours." He sighs. He studies them with a gentle smile, these new arrivals into his movable examining room. The mother is a big woman, with a forceful and anxious air. Her daughter sits quietly in her wheelchair, her square face and lank thatch of bowl-cut hair making her look vaguely medieval. Then he leans toward Jocelyn as if to say this part is just between the two of them. "Okay. Um. Why did you come? 'Cause Mom said you can?"

Jocelyn answers calmly, deliberately. "No, we found it on the Internet. And we just said, how are we going to go? And then we figured out how to get the money to pay for it. That's why we're here—because we found the money."

Kopits studies her. He's a tall man with a slight European accent and a reverent innocence that reminds me of Mr. Rogers, the kids' TV host. "That's wonderful," he says. "That's wonderful." Then he nods down at Jocelyn in that intimate just-between-us way and tries again. "Do you think that's a good thing?"

"Yes," Jocelyn answers in her firm voice. "It's a very good thing."

At last he's satisfied and turns to Evelyn. "Tell me, how is John Rogers?"

"John Rogers?"

"Dr. Rogers."

"I don't know," she says. "Is he in one of the Sydney hospitals?"

Kopits seems puzzled. As he understands the case, the Powells were having trouble getting adequate treatment in Australia. So how is it they haven't come across a prominent specialist like Dr. John Rogers? He files away the odd detail and asks Evelyn about her pregnancy. She tells him that she spent the last two months in a hospital bed because she kept going into labor every Friday night, and he smiles. "After the third time the doctor said, 'My night won't be ruined—you're going into the hospital!' "

They all laugh.

"Every Friday night from seven o'clock to eight o'clock, contractions one minute apart," she continues. "An irritable uterus. So she was born at thirty-six weeks."

"It wasn't really a normal pregnancy."

"No, far from it."

"And you had Nicholas subsequently?"

"Yes," Evelyn says.

"You are an optimist."

"We are," Evelyn says, laughing. "If it's going to happen, it will, but we'll make it."

Kopits smiles warmly. "That's good," he says. "I like that. Same here."

With that, Jocelyn and Evelyn begin really to relax. Kopits is so obviously painstaking, asking question after question about every phase of the birth, all in the same placid and gentle voice. He seems quite pleased by Evelyn's unhesitating, detailed answers as she tells the story of Jocelyn's birth and diagnosis. "You are a good historian," he says.

"We've been asked a few times," Evelyn says.

"And what happened after?"

"Once we got over the shock?"

He smiles gently. "After the shock."

"We brought her home."

Evelyn tells him about the terrible first year pretty much as she told it to me, except with more detail about the colds and runny noses and earaches. Jocelyn was a poor feeder. She vomited frequently. At eighteen months, she had arthritis so bad she required surgery. At twenty months, during treatment for her sixth or seventh ear infection, her eardrum was perforated by a careless doctor.

"She had stenosis since birth?" Kopits asks.

"No," Evelyn says. "She had a beautiful back until the doctor operated on her about four years ago."

Mildly, Kopits asks, "She had a straight back?"

"It was beautiful. It was something that we watched from the word go. Always made sure that she was sitting supported. We were very proud of that back."

As Evelyn describes the specifics of Jocelyn's spinal decompression, naming the four bones that were removed to ease the pressure on her spine, Kopits begins to shake his head. "Please don't shake your head," Evelyn says. "I'm already upset enough about it."

"Taking the bone out is not the problem," Kopits says. "The levels are the problem."

That's all it takes to nudge Evelyn to the edge. "From when she was very young she *never* walked well. She *always* complained of ankle pain, *always* complained of knee pain, *always* complained of leg pain, and whenever I saw the specialists, they would say 'Ah yeah, that's normal, ah yeah, that's fine, ah yeah, good.' And when she was about ten, she just refused to walk. The pain was too great. And then we started talking about spinal stenosis. Up until that time nobody had mentioned that word to me at all. From ten until thirteen, we would go back to the specialists for periodic visits. But no one would say, 'Watch out for this, do this, don't do that.' "

"There was a delayed decision in surgery?"

"Oh yeah, delayed. There was no discussion about surgery. When she was thirteen, all of a sudden they started realizing she was experiencing symptoms of spinal stenosis, serious ones. They kept checking her reflexes on her stomach and saying 'Oh that's not too bad.' In January, I said to them, 'She is not well. This just can't go on. This is not right.' And they said her stomach muscles don't seem to be reacting as they should be—come back and see us in six months."

Kopits seems appalled. "Acch," he says.

"The only thing we want you to watch out for is if she loses bladder control."

"Oh my God."

Evelyn describes their visits to specialists who kept saying Jocelyn wasn't doing so bad—right up until the day they were driving home from the neurologist's office and Jocelyn announced that she'd just wet the seat. They turned around and went back to the hospital, and the next morning they did the surgery.

"As an emergency."

"As an emergency. And then they informed me afterward, 'In hindsight, we left you a little long.' Yeah, in hindsight."

Kopits seems sad and a bit wearied by this story, as if he's seen this sort of thing so many times that each new telling leaches away a little more of his hope. "I think that, of all the lessons we ever learned, the most important one is that you cannot do limited decompressions," he says. "You go all the way to the sacrum and decompress everything. If not, you have a window edema. In other words, in a completely tight system you have opened a window, and then the cord will begin to bulge, which is going to give you neurological damage above and below. And then, full paralysis, or close to it."

Evelyn and Jocelyn look very somber. Kopits pauses, then continues in a softer voice. "This is how I saw the patients when I got into this field nearly thirty years ago. That was one of the reason that I set out to say, if we learn the natural history of what is going on, we can do the most benefit with the least amount of intervention. Because I saw terrible, terrible things."

Kopits gets out a new sheet of paper and asks about Jocelyn's physical therapy after the surgery. Again, Evelyn's frustration boils over. "I found it *very* frustrating," she says. "The physiologist didn't really have an idea what therapies to do. She'd do a little walking in the water, a little kicking. They didn't know what else to do. They did try getting a brace to try to fix the curvature in Jocelyn's back, but the doctor didn't even know how to get it on—he ended up hanging Jocelyn from the ceiling like a side of beef before they just gave up. The back was bent all the way to seventy-two degrees at one bad point."

Kopits looks appalled. "It was seventy-two degrees?"

Evelyn nods eagerly, so thrilled that somebody understands. The nurses kept trying to straighten Jocelyn's arm to take her blood pressure, even though she *told* them that dwarfs don't have full arm extension. She got so angry that finally she just stole Jocelyn out of the hospital and never went back. "From then to now, three years, we're not seeing any doctors, not seeing any physiotherapists. We're not doing anything."

"And what about pain?" Kopits asks.

"If she walks from our room to the lobby, by the time she's down to the restaurant she's ready to fall."

Kopits nods. "I saw her standing this morning."

"Did you watch the knees? When you see them bending, you know that she's ready to sit down—or fall down, whichever comes first."

Kopits shakes his head and makes a note in her file. "Now, going back to the present." He finds his place, then looks up at Jocelyn. "At this time, standing is a problem?"

"Yes, sometimes," Jocelyn says.

"And walking independently—do you use a walker?"

"No."

"So how far can you go? How many steps?"

"She walked from her room to here and that's it," Evelyn says.

Kopits does the mental calculation. "That's probably fifty, sixty meters."

"By that time she's dragging her right foot," Evelyn says.

"When you walk, you get heaviness, feel like falling?"

Jocelyn nods enthusiastically. Like her mother, she's visibly pleased that someone understands. "It starts in my toes—it goes pins and needles, numb, and then lead, heavy lead. First the right leg to the knee, and then the left leg starts." She seems neither bitter nor self-pitying, but there is a touch of dignified protest in the almost clinical way she speaks.

Kopits smiles. "You describe it well. I've seen it thousands of times." Then he turns back to his chart. "What is the distance at which this begins?"

"It depends on how my back is, how straight it is—if it's straight, probably twenty, thirty meters. If I'm lucky."

Kopits's face falls. "Oh," he says. Jocelyn and Evelyn exchange alarmed looks. "This leads me to believe that your space situation is extremely, uh, critical," he continues. "Normally, you see, the spinal cord moves within the spinal canal. When I stand up, my nerve roots go into the canal one inch. I sit down, they go out of the canal one inch. In you, no motion. The spinal cord is fixed to the canal. Now, imagine this is a tube." He holds his hands a foot apart. "I'm bending the tube. This is what's happening to you—your kyphosis actually *stretches* the size of the cord."

A somber pause follows.

"We haven't hit the best yet," Evelyn says.

"There's more?"

"This is the Powells you're talking to," Evelyn says, laughing bravely. "If it's going to go wrong, it will. On December 29, 1995, we were just thinking what a wonderful Christmas we'd had, how well she was—and she woke up and had one *massive* headache. And she stayed really flushed and a few days later—"

"Nausea," Kopits says. "Then migraines."

"And then loss of concentration for school." Eagerly, Evelyn tells of making one more visit to the neurologist (at his office, not the dreaded hospital) and blowing up at him and how he sent them to a psychiatrist who said that maybe the problem was stress or a "dominating mother." The frustration and anger rise in her voice. "She was in so much pain. We were losing the only quality of life we had. From the neck down, we can deal with disabilities, but once her brain is affected, and her schoolwork is affected, there goes her future. She wants to be an accountant. She's above average in her schoolwork, she's a very clever young lady. And I was angry that this was slipping away from her and they weren't prepared to do anything. That's when I got in touch with you. Do you remember that phone call in July? That was two o'clock in the morning to us. I couldn't wait for five o'clock. I woke her up and said, 'I just talked to Dr. Kopits in America, and guess what, there's something wrong.' 'So it's not me?' I said, 'No darling, it's not you.' We were both crying, we were so happy that there was something *wrong!* While they were trying to convince us that it was psychiatric!"

"Oh my God," Kopits says.

She mimics the doctors' voices. " 'We've seen you so often, and your mother is so overbearing' and all this sort of rubbish. 'Your mother is a tiger!' And yes, I *am* a tiger. Why do I have to fight to get information? Why do I have to *become* like this?"

She's on the verge of tears. Kopits nods encouragement.

"And then I rang the doctor and I told her what you'd said. And I got a ticking off that I had rung a doctor in America. Then she said, 'We knew that. We were pretty sure it was intermittent hydrocephalus.' Well, I was devastated. 'Why didn't you tell me!?' "

"Oh my God," Kopits says again.

"And she said, 'Oh, because we're not convinced.' And I said, 'Well, what will it take to convince you?' She said, 'Well, we'll see how much worse she gets.' "

Kopits moans.

That was when they turned to a chiropractor. He began treating her with "cranial motion," a heavy head massage.

"Tell me—isn't there a neurosurgeon there?"

"Yes, but they wouldn't do the intercranial pressure. They refused to do it."

"They refused to do intercranial pressure?"

"They said it was too dangerous."

"Too dangerous?"

"I said, 'I'm not stupid, I know it's dangerous. . . .' "

"You can lose a brain to this—"

"I know!"

"And go blind," Jocelyn adds unemotionally, as if she's just indulging a penchant for precision.

"What I'd like to know is, where do we go from here?" Evelyn pleads. "What are we looking at? I'm—I'm—I'm lost. I'm frustrated. I'm a mother. I'm angry."

Kopits gives her the kindest look I have ever seen. "You can't go back like this," he says.

"Oh, I know."

"She has to be treated. The question is, Where do you find the money to do it?"

Evelyn laughs nervously. "Yeah, exactly."

"I already have the neurosurgeon for you. He's extremely talented."

Suddenly both Evelyn and Jocelyn are whimpering quietly, tears leaking down their faces.

"You'll do it," he says. "You'll do it. This is going to be solved, and you are going to be well."

Evelyn turns to Jocelyn. "Are you okay?"

Jocelyn nods, sniffs.

"This is sheer nonsense, what has happened to you," Kopits says. "This cannot continue."

"I know," Jocelyn says.

"You are too precious."

It's getting very emotional in this room. Kopits launches into an exhaustive monologue on the medical risks and benefits, explaining it to them in vastly more detail than they could possibly need, clearly trying to soothe them with his knowledge and his voice and his patience.

"Little thing, though," Evelyn finally says, sniffling. "Where do we get all the money?"

"A quarter of a million dollars, I think."

"Getting the five thousand to come here was hard enough."

Then Kopits turns to me. "You have a lot of money," he says. "You can make it happen."

We all laugh, Jocelyn and Evelyn and I, assuming that this is a joke, that in a moment Kopits will drop the solemn tone and laugh along with us. But he doesn't. He continues to stare at me in a strange way, as if he's snagged me in some kind of *Star Trek* tractor beam. "I wish that were true," I say, a little nervously.

He keeps boring in with that stare. "You see, you're in a very delicate situation," he says. "Because you are getting hooked. You are getting a part of the moral responsibility that all of us have to help her." He lets another long pause stretch out, and then he finally cracks the smile we've been waiting for. "Let's put that on your back," he says.

"Thanks a lot," I say, relieved.

But then he lets the smile fade and keeps boring in on me with that solemn intensity. "You needed this," he says. And the moment drags on until it becomes strangely unsettling, rattling me in ways that I understand only much later. Part of it is Kopits's moral authority. Already I know that he sacrificed a marriage and family and a rising career at Johns Hopkins Hospital to treat his dwarf patients night and day, that he's famous for the care and time he lavishes on them, that the big career-making book he's been talking about for years never quite seems to get written, but his dwarfs love him with rare devotion. I'm also rattled by all the things I've seen these last two days, the kids zipping around in their wheelchairs and the scars on the legs and the little digs like Michael's comment, that first day at the track, about how differently women would treat us if we went into a bar together. And there's also the thought of my mother in her hospital bed, recovering from cancer surgery. In so many ways, I'm off balance. So I look back in Kopits's solemn eyes and say, "Yes, I believe I did."

For one more moment he stares, locking me in that tractor beam, putting his moral mojo on me . . . and then finally he relaxes and smiles, looking from face to face. "It's remarkable how this happened," he says. "How the right people are in the right place when it matters. This is what you see here, a true miracle."

Then he glances down at Jocelyn's medical record. When he looks up, there is an expression on his face so eager and cheerful it is close to mischief. "Now . . ." he says, picking up his pen and playfully stretching out the pause. "There is more to come!"

Three

BACK IN MY HOTEL ROOM THE AIR CONDITIONER HUMS and the bedspread is turned back and everything is reassuringly antiseptic and I can't sleep, I'm so buffeted by images: Kay's sad eyes! Meredith with her silver cane! The tricycle baby! Squat Jocelyn and her giant mother! And these pseudos with their big heads like beautiful prows stuck onto their little rowboat bodies, magical in the way that stage sets are magical when you see the raw two-by-fours behind the gloriously painted flats. That's what actors and big-eyed aliens and babies have in common, big heads on little bodies, and there's got to be a reason for it—like those big mysterious heads on Easter Island with no bodies at all. And I'm buffeted too by all the emotion, by Michael's doubt and Evelyn's fear and Kopits's concern. Unable to sleep, I toss back the covers and get dressed. It's only midnight, might as well check the lobby.

Walking toward the ballroom, I realize I'm going to have a problem when it comes time to write all this. As I know already from skimming the

Internet, there's only one acceptable dwarf story in contemporary journalism and that's some variation on the theme of "Big Hearts in Little Bodies." Here are headlines picked almost at random: "Gutsy Pete Moore Is Barely 27 Inches Tall," "3-Foot Twins Are Big Men at Selling Real Estate," "She's 4′2″, But Amy Roloff Has Never Let Her Size Stop Her from Getting a Lot Out of Life." The headline "Ordinary People" gets used a lot, with subheads like "Dwarfs Exchange Vows 'Like Everyone Else' " and "The Giebs Pursue Regular Existences Just Like Those Folks in What They Call the 'Tall World.' " The LPA convention is invariably celebrated as "the first place we can be comfortable." The courage theme brings out violins and lyric flights. One columnist writes:

I wish I could describe Danny Singer's smile. It's the smile of a kid who has broken more than 50 bones in 13 short years, without breaking his spirit. It is a smile which is remarkable because it is there at all. I wish I could describe Phyllis Penton's tiny handshake—the handshake of a 36-year-old woman who is not quite 3 feet tall, but who has put herself through college, worked as a guidance counselor and who has, according to her doctor, twice 'come as close to dying as anyone can come and still survive.' I wish I could describe these things. But who has the word to describe miracles?

The only permissible variations seem to be tales of pure science, and even those tend to mutate into upbeat stories like "A Doctor Fights for the Little People." Later I will find academic books and articles in specialized journals with material on the shock dwarfs feel at entering the LPA lobby, or their interesting tendency to "present" cheerful selves and then withdraw, or their complex process of barrier formation, but even the best of these seem to have a reflexive need to parrot the "nothing special here" party line. "Dwarfs are ordinary people who bear the burden of a physical difference," writes Joan Ablon, author of an otherwise intelligent sociological study called *Little People in America*, before devoting the next three hundred pages to describing exactly how dwarfs *aren't* like ordinary people. Eventually this need to dismiss begins to seem like more of a symptom than an argument, and I find myself saying, wait a minute—you're telling me that living in this body doesn't affect you?

The body you're born with? The face you see in the mirror? Like we're all these pure floating souls communing across the ether? I admire the democratic spirit as much as the next guy and I really don't want to disappoint Dr. Kopits, but I'm not going to delude myself.

When I get to the ballroom, I see the tall girl dancing with that same little guy. Now that I'm a little more dwarf-savvy I can see that he's probably a pseudo because he has a normal-looking face—no pushed-in nose, no bulging forehead. He's also taller than most of the other dwarfs, right near the four-ten limit. When I get up my nerve and ask them for an interview, they agree without hesitation. In fact, they're thrilled. Their names are Chris and Emily, and they met at last year's convention. They've been dating for a year. The hardest thing is living so far apart, Chris in England and Emily in Minnesota. And the age difference. He's twenty-two and she's sixteen. They've never even talked about the height difference before and it's kind of embarrassing, but exciting too. Her parents had a real hard time with it, Emily says. "They're here, you can talk to them." And some of her friends did too. "They would say, 'You're dating *what?*' and I would say, 'A dwarf,' and it would be, 'That's kind of weird.' They wouldn't say it to my face, but I could see it in their eyes." But once they got to know him, they liked him. And her little brother loves Chris to death, always wants to tag along. He's a dwarf too. That's why they come to these conventions. She gets very offended when people call him names or say, "Oh, your brother's a midget." So how could she refuse to date a dwarf when her brother is a dwarf? "You have to give everybody a fair chance," she says.

Later, I talk to Emily's father. He's about six feet two, a dentist from the plains of North Dakota, with a shy and gentle manner. It's been good for the family to come to the LPA conventions, he tells me, good for Emily and good for his son too. When his wife gave birth to their son, they didn't know anything about dwarfs or little people. It was very, very scary. But once they got in touch with the LPA, the support was so great and John made friends all over the country and he got a chance to compete in the athletic events and they've met so many terrific parents and Emily too, even though she's an average-size person, she's made some wonderful little people friends. . . .

"How did you feel when she started dating Chris?" I ask.

His voice gets a little fluttery. "It was . . . another experience in life and . . .

it was gratifying to me to see her enjoying herself. She was real happy and having a good time. I was a little more concerned about the age difference, rather than the height difference."

He goes on about this for a while, the age difference and Chris coming to visit them in Fargo after last year's convention, and no matter how much they liked him, and they did, they couldn't help thinking about how much older he was.

Earlier one of the dwarf girls told me, with a hint of bitterness, that Emily had dated another dwarf before Chris. Just to see what he'll say, I ask Emily's dad if he knows anything about that.

He answers a bit mournfully. "I really don't know the exact nature of the relationships that Emily's had with little people," he says. "But I know she's had some . . . wonderful social activities."

"Do you have any queasy feelings at all?"

"Yeah, yeah, to some extent," he says, taking longer and longer pauses as he goes on. "But now, today, it's like . . . anything goes. And you know, once we . . . came to grips with the idea that our son was a little person, a dwarf, then it didn't . . . you become familiar with that type of condition, if you will. I remember it was devastating at first, but now we accept it as . . . part of life. And once we accepted the fact that he is human, he is a human being just like the rest of us, I think it helped me get through the idea that my daughter wants to form a relationship with a person of short stature."

Humbled, I shut up.

"I guess I haven't really dwelled or thought about it that much," he continues. "The queasiness you ask about, I think, sure, we all have our visions or our ideals of who we'd like our daughter to be with. It might cut into my wife's ideals a bit if it became a long-term relationship or a marriage, but as long as she's happy and enjoying herself and successful in life, that's more important to me than the size of the person she dates."

As we're speaking, Emily comes up and throws her arms around him. He gives her a shy and happy smile. When she dashes back onto the dance floor, his eyes follow her.

. . .

For lunch today Michael and Meredith go to Bennigan's, driving in a car Michael rented so he could *get away from all the friggin' little people.* Right now, they're his biggest problem, not only the masses of them reminding her that she's a dwarf and he's a dwarf and the cliques that are watching him and whispering snide things and the other guys who keep hitting on her. Like his friend and roommate, David Brookfield. Maybe even his pal Gibson Reynolds III, an ironic, urbane dwarf from Connecticut who seems very much to the deck shoe born. And he wants her to know, this is what it's like to drive a car with a little man. This is what it's like to go to a restaurant with a little man. . . .

As they walk in the door, Meredith turns to him. "This is a big step for me," she says.

"How do you feel?" he asks.

She reaches over and puts her hand on his. "Nice," she says. "I like it."

But Michael needs more. What about the other guys? He knows they've been asking if she's with him. He knows she's been saying no. But why? *Why?* "I'd like you to say, 'Kind of, yes,' " he says.

"I don't know," Meredith says. "I just . . . I don't know."

Then Michael heads to the gym with his friend Kay and plays a ferocious game of basketball. But his mood is sour, and when it's over he collects his medal and Kay and rushes for the exit. A blind man could read the expression on his face—somewhere, this minute, this second, some other guy is almost certainly hitting on Meredith.

Then he stops. Standing in the hallway by the door is a tiny, tiny woman. She's at least a foot, maybe a foot and a half, shorter than he is, wearing a strand of fat pearls.

Michael squats down to her level. "I just wanted to let you know," he says, "that time in the chatroom, when you jumped on me, I thought that was very unfair."

"Well, sorry," the tiny woman answers.

"I mean, I don't even know you. Why would you do that?"

The woman shrugs, not giving an inch. "It had to be done," she says.

Later, Michael tells me the story. He was in AOL's little people chatroom with a bunch of his friends and Carrie, the girl he went on-line shopping with, the one who would never send him a picture, she and his friend Kris started teasing another woman about something or other. Then Martha jumped in. That's her name, the woman with the pearls, Martha Holland. All Kristopher ever did was go after the pretty girls, she said, ignoring anyone who was different or less attractive—or extra short. And when Michael tried to defend Kris, she shot off a message saying he wasn't any better.

At which point Carrie pipes up. What do you mean, Martha?

And Martha poured it out, trashing Michael and Kris and all the dwarf men who would go after only pretty women. It wasn't right. It was cruel. It made being in the LPA just as bad as being in the tall world. "I've waited a long time to get this off my chest," she said. And Carrie heard Martha's anger and felt her pain and never spoke to him again.

As Kay and I step to the side, Michael continues to talk to Martha in the same firm tone of voice, pleading for understanding and also arguing a case. He's speaking so softly it's hard to follow, but I can hear her arguing back. "No, it's cruel," she says, her voice bouncing off the tile gym walls.

"You messed me up with someone I really cared about," Michael says.

"Then maybe she's, you know, mistaken."

Michael keeps arguing with her. Finally she says, "What's done is done."

"I'm sorry you feel that way," Michael says.

. . .

Sitting in the lobby, Kay says yeah, she was there in the chatroom that day with Carrie, and it's true that Martha seemed to get quite a thrill out of sinking her stinger into Michael. Even today, she still seemed so proud of the power she has, the power to hurt. But it's so much more difficult for the small ones, the diastrophics and SEDs. Even in the little world, it's hard to be smaller. "As much as I love Michael as a friend," Kay says, "I grieve for her. When she looks at me, she sees a certain person, and she won't give me a chance. But I was on both sides. I was on the ugly—I was on the unattractive side, and I was having difficulties walking and stuff, and now . . ."

She lets it trail off.

She lost seventy-seven pounds, but not the memory of those seventy-seven pounds. "I worked hard to get where I am," she says, sighing. "But I always will sympathize with her. I don't think she had the right to do what she did, but I can see her anger."

. . .

I just realized that I've been here three days and I still haven't talked to a single one of the really tiny or misshapen dwarfs, the diastrophics and SEDs and the others. They just seemed so damn different—unapproachable, like foreign countries with harsh, impossible languages. Instead, I've been hanging out with Evelyn and the "mainstream" dwarfs, the achondroplastics—and, I must admit, not the old ones or the teenage ones either. I seem to be more comfortable spending time with the ones who are more or less my age, the ones who look less dwarfy. The better-looking ones.

When Kay leaves, I mention this to Gibson Reynolds, another good-looking achondroplastic. "When you're an achon," he tells me, "it's almost like saying you're glad you're white: 'Oh, at least I'm an achon.' "

I am going to have to talk to Martha.

. . .

At the talent show, I find myself sitting next to a little woman I will call Andrea. She's about forty, with frizzy hair and one of those square achon faces that's heavy in the jaw. I met her before in the lobby and we quickly fell into a teasing kind of relationship, the main joke being that I'm bent on asking questions and she's intent on staying out of the public eye. "So what brings you to the convention?" I ask now, putting on an ironic mock-interview voice and holding out my tape recorder. "Are you looking for a love connection?"

She answers dryly. "No, I'm looking for sex."

We laugh and go back to watching the show. Up on stage, a girl in a bouffant cut and a white sleeveless gown starts singing a song from *My Fair Lady*. Then three cute little dwarf teenagers in cutoff jeans and white tops sing "Girls Just Want to Have Fun."

Then Andrea leans over again. "I was only partly joking. Because the convention is too short to find love, really. But sex . . ."

I hold out my tape recorder and pretend to be very somber. "What are your chances of getting lucky?"

As I ask the question, I can't help checking her out more closely. She's wearing a sleeveless, loose-fitting dress, has her hair tied in a loose ponytail and doesn't shave under her arms. Her jaw is very square.

"I'd say they're pretty darn good here," she answers. She looks around, thinks, then nods. "Yeah, I could be lucky if I wanted to be."

Onstage, a dwarf kid plays a really nice version of "Fantasia."

At that point, Andrea insists again that she doesn't want to be written about, so I promise about five times not to use her name and finally she continues, talking into my tape recorder. "Anybody could be lucky here, pretty much. That's not boasting. I think people want to be loved, and they take more chances here. They aren't going to take them in the average-size world, usually. I mean, I don't pick men up in bars in the average-size world."

"But here?"

"What I meant—what I mean is—I'm a—stop!" She laughs.

"But I need the aggressive woman's point of view."

"Oh, great," she says.

Onstage, a kid wearing a big hat does a professional-kid-entertainer version of "Let a Smile Be Your Umbrella," dancing smoothly and spreading his little arms for the big notes.

"You don't necessarily have a lot in common with people, just 'cause they're short," Andrea continues. "That's the main reason I don't come here looking for love. But to be honest, there is a part of me—I mean, if there weren't the diseases—I would probably come here looking for sex. You know, you have to be practical." She covers her mouth. "I can't believe I said that."

Then I'm distracted by a toddler who is trying to dance in his custom-made backward walker—instead of leaning forward like an old person, he drags his walker along behind him and leans back against two foam-rubber handles. He's one of the extreme ones. "Poor little guy," I say.

Andrea gives me a sharp look. "Why do you say that?"

"I don't know . . . he's so little, and he's got so many troubles."

"But he seems happy, doesn't he?"

The little boy's mother kneels to take a picture. He leans into the foam rubber handles and puffs out his chest.

"Yes," I admit.

"So don't pity him," she says. "It doesn't help him."

. . .

Ten rows up, Evelyn can't keep her mind on the show. This morning Dr. Kopits called and told her to fly Jocelyn up to Baltimore for an emergency consultation. The threat of paralysis is that near, he said. She put down the phone and said to Jocelyn, "How are we going to do this? How are we going to do this?' The ticket's $499 one way, for one person. So we're talking $2,000 to go back and forth to Baltimore. I don't have that. I'm lost. I have no resources."

Then she called her husband in Australia and told him what Dr. Kopits said, and he freaked. Later she writes in her laptop:

As I expected he had 1,000 questions and no way was there a shred of trust. 'Do they know what they're doing?' 'Have they done this before?' 'How is her back going to be supported?' 'Who is this guy?' etc. etc. I'm busy organizing stuff and as usual David wants to wait!

And the thing is, David didn't know what they had *seen* here. He hadn't seen the packs of achon teenagers roaming around, hadn't seen their boisterous chumminess, hadn't seen how sophisticated they all are and how comfortable they are in their skins. They *have* something, these American dwarf teenagers, something that Jocelyn doesn't have. And maybe David wouldn't even appreciate it if he did see it. He's so stiff and restrained, not the kind of person to value that sort of thing. And look where trusting the doctors back home got them! Look where it got Jocelyn!

Help!!!!

And now Jocelyn is sitting next to her at the talent show clutching the book *Healing Hands* that Bunny Phipps gave her about the famous Dr. Ben Carson and how he rose from the ghetto to the top of the pediatric orthopedic field and led a team of forty doctors and nurses to separate a pair of Siamese twins and if he operates on Jocelyn it's going to cost *a quarter of a million dollars* and Barry Phipps keeps saying, "Don't worry, don't worry." But she can't *help* worrying!

Help!!!!

And then the show is over and there's Barry over by the door saying "Don't worry" one more time. "We'll get her up there even if we have to rent a car and drive eleven hours up to Baltimore," he says. "We've all adopted her now. She's not only yours, she's ours."

Evelyn laughs, dazed and a bit giddy. "That's all right. I'm happy to share."

People are folding chairs and cleaning away props. Barry goes off and Evelyn turns to me. "I wasn't ready for this," she says. "I thought he was going to say, 'Be prepared for this, this could happen in the future—next month, next year.' So this completely overwhelms me. I'm scared."

She tells me about the phone call to David and how he insisted she fly home right away, more understanding this time than when she was tapping her private thoughts into the computer. "He's just as scared as I am," she says. "Once you have a bad surgery, you get scared to go the second round. Is this going to work? What are the complications involved? You're thousands of miles away and she's in this place I can't imagine talking about these people I can't imagine, and now she's telling me that she's going to stay another eight weeks and it's going to cost two hundred and fifty thousand dollars."

The words are pouring out of her, gushing and bubbling until Jocelyn goes off with Barry and Bunny and the room is empty. There's a fast in-the-trenches friendship sparking between us, fellow tall people and caregivers, and I love her when she makes me swear never to use the words "silver lining" in anything I write. "I hate it when people say they don't even acknowledge their children have a disability," she says. "Come to our house and there *is* a disability there." As I tell her about my mother and some of the things I've been through this terrible cancer summer, about my father's suddenly advanced age and increasing weakness, she comes back with more stories of their past and how scared Jocelyn is—"really scared, enormously scared, scared with capital letters, underlined, SCARED"—and how much comfort Bunny is giving her and how some people at the convention say that Dr. Kopits is "surgery-happy" but others say he's the greatest doctor in the world and she's just going on her instincts and she *thinks* he's a good man and she thinks Baltimore is the right thing. . . .

Help!!!

On the verge of tears, she distracts herself with another clear-eyed observation. "I never had little people as friends before, and having seen them in such a large group, it's still hitting me that they are human *beings*, with *lives.* And I have a daughter who is a little person, and I am fighting for her to have a life, to have a career, to be in love, to—to—to do all the things that the others just *do.* And yet, I watch them go past, and I think, 'That's a human being.' "

. . .

By Wednesday, Meredith is more relaxed—a bit. "I want to go back to my world," she moans. "This is not my world."

She's gone to some of the workshops, attended events put on by the Dwarf Athletic Association of America, done some shopping and gone on some of the special tours. But, well, she's having trouble handling all the eager men, for one thing. "It's getting almost a little bit exhausting. At the same time, to be perfectly honest, it's a dream come true. When I was a teenager, my best friend Alison was really beautiful, and she always got the guy and I was always the friend. Guys would be like, 'You're really pretty, but you're like my sister.' Now I'm the one in the spotlight. I feel good about myself here, sometimes. But then I feel like, Why can't this happen at home?"

Even talking eye-to-eye isn't the unmixed treat she thought it would be. "It's almost a little uncomfortable. I'm so used to people bending down, I don't know how to handle it when I'm standing there with Michael and I'm talking face-to-face with him."

All week long, it's been back and forth like that. She really likes the people, some of them. She identifies and feels understood and accepted and all that stuff. But then she recoils. She hates the idea that this is where she's *supposed* to belong. One thing for sure, she'll never be one of those people who looks forward to this all year. "I'm never going to really fit in like that," she says. But I guess everyone else, because I look like them, thinks that I fit."

I ask her why she came. Was she reaching out to the Little World?

"Absolutely," she answers. "I really came here on like a vision quest, a search for identity—to see what it's like to talk face-to-face, eye-to-eye with someone."

And did part of that quest involve a small-statured man?

She squirms. But, like almost all the little people, she wants to be understood, so she's startlingly open to the rudest questions. "I don't know. I'm twenty-two. He's much older than I am—thirty-four. He wants a relationship. I don't."

He also wants children. She's not so sure. "I've had about fourteen multiple surgeries, and I deal with pain every day," she says. "I would never want to do that to a child."

Then she mentions that she's met another guy here she's attracted to—an average-size guy. "I feel almost disappointed in myself that I've come all this way, to pick an average-size guy," she says. "And I have to wonder if that says something about accepting myself. Why do I shy away from short-statured men? It's not that I don't see them as men, but . . . I'm *attracted* to a tall man. It's something about them that's protecting, and safe. I want someone who's going to take care of me, someone who can physically—I don't know, it's just something about a tall man, with a long back, and long arms. . . ."

And she felt weird at Bennigan's—people were looking at them like, "Oh, how cute," and she *hates* that. But how would Michael take it if she dumped him? He's so considerate, so caring, so . . . small. "I guess I feel like a taller guy can take it."

God! The whole thing is just *exhausting*. "I need to sleep for like a week," she says. "For some reason, this is draining the life out of me."

• • •

That night at the dance, I find Andrea sitting at one of the long tables where people park their drinks and hopes. We make small talk for a while but something seems wrong.

"You seem sad tonight," I say.

She shrugs. "I was crying earlier," she admits.

I ask her why and she shakes her head, giving me a look. "It doesn't have anything to do with being short," she says.

"Tell me anyway," I say.

She checks my tape recorder to make sure it's off and then tells me that her father was very demanding and critical. He was a scientist and wanted her to be a scientist too, and she just wasn't interested. He never got over that. But he was disappointed in all his kids, the tall ones too. When one of the sisters came out as a lesbian, he completely stopped talking to her and even cut her out of his will. So Andrea hasn't spoken to him for years, and now he's sick and she's trying to decide if she should go visit him. Then it's my turn, and I talk about my critical father and sick mother. Every now and then someone comes over to the tables and asks someone to dance, and it occurs to me that asking Andrea to dance would be the decent thing to do. But frankly, I don't want to—how would I do it, on my knees? Bending over? Trying not to bend over? Anyway, I'm enjoying the conversation. It's a relief to be off the subject of short and tall, to be talking about the ordinary human stuff that we have in common instead of the stuff that separates us.

Then the lightbulb goes off. "I have to write about this," I say.

"I *don't* want to be written about," Andrea says.

"But it'll make the story richer."

"But this has nothing to do with dwarfism."

"That's *why* I have to write about it."

Then Andrea gets upset. *Very* upset. She accuses me of stealing an interview, of breaking my promise not to quote her. I point out that the tape recorder isn't running and that I'm asking for permission, but it does no good. "I told you," she says, "I do not want to be written about. I was very clear about that."

A little panicked myself, I do my best to calm her down. I swear up down and sideways that I would never quote her or anyone else without permission, which is true. Finally, she does relax a bit. "Anyway, it's just typical family issues. Although—oh, here's a good one, you'll like this." She pats my arm. "My grandmother once said to me, 'I've got a good idea for you, you could stay home and be a typist, that way nobody would see you.' "

I wince, but Andrea just smiles, like it's some kind of triumph. "And my dad got really mad at her," she says.

"Well, that's good. He took your side."

"I'm sure it wasn't because he thought it was mean. He just didn't want me to be a typist."

. . .

An hour later, well after midnight, Andrea finds me. "I can't believe I told you all that stuff about my father. You promise you won't quote me?"

Needled by her distrust, I decide to torture her a little. "Did you say off the record? I don't remember you saying that."

"I'll sue you."

"You can't sue me."

Her face cocks back in sudden horror. It's as if all her worst fears have been confirmed, not just about me but about everything. It happens so fast I don't have time to switch out of teasing mode. "What can I tell you?" I say. "Seduction and betrayal, that's the gig. I worm out your most intimate personal secrets and then expose them in the national press."

I don't think I've ever really seen a face "fall" before, but Andrea's face falls from ten stories. She wheels around and stalks for the door.

"I'm just *kidding*," I say. "Wait up!"

She ignores me, cutting a sharp angle through the lobby to the elevator. Despite the hour, couples are still sunk in the sofas, and teenage girls walk arm in arm. They give us curious looks.

At the elevator, Andrea punches the UP button like it might be my OFF button instead. "I can't believe you're this upset," I say. "Just think it through logically for a second—if I was planning to betray you, would I announce it? I was *joking*. Anyway, there's a thousand dwarfs in this hotel. I don't need to steal an interview with you."

She won't look at me.

"You're *way* overreacting."

The elevator comes and she gets on, looking at the floor.

. . .

Tuesday starts with a call from Baltimore—Dr. Carson can fit them in on Friday morning. Evelyn rings off, takes a deep breath, and dials the numbers for Australia. Then she hands the phone to Jocelyn. "You talk to him first," she says.

When David answers, Jocelyn starts telling him about all the dwarfs. The thing is, they're all in different *stages.* Which is so strange because until this week she thought all dwarfs went into wheelchairs when they were ten. Maybe that's because the only place she saw them was in hospitals. She never saw them walking around like the kids here, just going and going and going.

Then Evelyn takes the phone and tells David she's not going to rush into anything, not to worry so much, just trust her a little bit.

Well, David says . . . if it really is an emergency . . . Then it's lunch and afterwards Jocelyn lingers downstairs, sitting in her wheelchair at the end of a long hall lined with a series of tables where people are selling specialized golf clubs and bicycles and step-doublers and other dwarf-enabling products. In her lap, she holds a book called *Dwarfs Don't Live in Doll Houses,* the story of a small woman who overcame many obstacles to become one of the first dwarf lawyers. As she looks over the bustling scene in front of her, imperturbable as a little stone idol, I ask her once again how she got to be so sure of herself, so eerily *solid.*

"I don't know," she answers. "I just came to it very quickly. I know that I'm Jocelyn. I know that I'm sixteen years old, and I know I come from Australia. I'm proud to be a dwarf. I'm just me, a smaller package than everybody else."

But how did she reach that place?

"I don't know," she says again. "I see a lot of people who are still searching for who they are, and this amazes me, because I did that quite a few years ago. I can't relate to sixteen-year-olds anymore, because I've come through so much, and they've gone through nothing, compared to what I have. Some of the things they go through to try and find themselves, it's just ridiculous to me. My brother is going through it now—he's fourteen, and he can't wear hats now. He used to wear the ones that come down and protect your eyes, but now he can't wear those, he can only wear the baseball caps. He shaved his head at one stage. Then he had to wear shorts, because it wasn't cool to wear pants."

"But aren't kids supposed to be trying different things? Exploring the world and all?"

Jocelyn shrugs. "I can't change, so why bother?"

. . .

In the lobby, I run into Barry Phipps. He tells me they're raising money to get Jocelyn up to Baltimore and they're going to present them with tickets later, after the next event. I give him a hundred bucks—from my own money, not my expense fund. That seems important.

. . .

Late in the afternoon, it's time for the LPA fashion show. I run into Jocelyn in the ballroom and this time when I ask her how she's doing, she finally admits to a recognizable human emotion—she's scared. Very scared, in fact. About the operation. She was the one who swore off hospitals after the last operation, she informs me. Her parents just went along with it. Then when her back started getting so curved and the headaches got so bad and she started thinking about hospitals again, her parents had flipped to the other side and weren't so sure it was a good idea. "They were like, okay, you can do that if you really want, but not until you're eighteen." But seeing the dwarfs here really has had a big impact, and she trusts Dr. Kopits more than she's ever trusted a doctor. He really seems to know what he's talking about, and he understands her symptoms so well. And he really seems to care. "In Australia they just wanted me to get lost. They wanted me to go home and forget all my symptoms, because they didn't know what to do."

Despite the content of what she's saying, Jocelyn uses her usual flat voice, as if she's just stating the facts. This time, she's not going to let them touch her until she knows every detail, she says. She's going to talk to people who have had it done. She's going to ask about the fatality rate. There will be no shocks and surprises. "I wasn't ready last time. And they didn't explain at all about the operation. My mom didn't tell me I might be paralyzed, and it wasn't Mom's fault, because she didn't know how to tell me and she was so shocked. By the time she got over it, it was three months after the operation. Even then, she was afraid to tell me."

She might as well be reading from a computer instruction manual. "I'm almost eighteen," she drones. "I can make decisions myself."

Again I feel it, her unnatural force. She's so *compressed.*

When I mention it, she doesn't blink an eye. "I've always been stuck and not been able to go anywhere," she explains. "I've always had to be carried. I thought I'd never get much better. That's why I got very emotional on Sunday when Dr. Kopits told me."

I thought she was crying out of fear, I tell her.

"No, happiness," she says.

. . .

On stage, a dwarf girl in a black cocktail dress strikes poses. "Jennifer is from Rhode Island. She is seventeen years old. She loves going to the beach and just having fun."

Oohs and ahs and a big hand.

Then it's a toddler in white shirt and dark suspenders, an Indian woman in traditional Indian dress, three small women in gowns and three more in miniature kimonos. "This shows we have brothers and sisters in the LPA everywhere," the announcer says.

Evelyn sits in a row near the front, wishing Alecia and Nicholas were here to see this. Everyone is so cynical these days. The modern world is so anti-family. That's why she doesn't let her kids watch *The Simpsons.* It doesn't respect parents; it teaches kids to be bratty.

A dwarf Santa Claus takes the stage and the invisible announcer tells us that when he's not delivering toys, "He fixes wheelchairs and crutches and scooters for people who can't afford to get them fixed."

Evelyn smiles.

Then a personal note from the announcer: "I've just been speaking with a lady from Kansas who says she hasn't been asked to dance, so, gentlemen, you go out there and you ask every lady here to dance. Doesn't matter who you dance with, just so long as you go home with the one that brung ya!"

Evelyn smiles again.

After the show, Barry and a small group from the Atlanta LPA take Evelyn and Jocelyn aside. "The committee wants to give you something," one says. Then Barry hands over a little bag and Evelyn reaches in and pulls out a yellow LPA T-shirt. Inside the T-shirt is tissue paper, and wrapped inside are the two round-trip tickets to Baltimore.

Evelyn starts crying. Then everyone starts crying.

"You guys are my family now," Evelyn says.

. . .

At the airport Wednesday morning, Michael is gazing into Meredith's eyes with an intensity that almost vibrates. So Meredith's mom and I go across the room to give them some privacy. After a moment, I see Meredith drop her head onto his shoulder. Michael lifts her fingers to his lips. More gazing ensues. Then I can't stand it anymore, so I walk over and hand them my tape recorder. "Turn it on if you feel like it," I say. Back at my seat, I watch Michael take out a gift bag. Meredith opens it with a smile.

Later I listen to the tape. "This is the first actual present I've ever bought you," Michael says. "We've had a lot of firsts—first lunch, first kiss. . . ."

"Don't say that on the tape!"

"Okay," Michael says. "Scratch that comment about 'first lunch.' "

Meredith laughs. "That was really funny!"

She reaches into the bag and takes out a teddy bear. "Aww, it's so *cuuuute!*"

"It's Winnie the Pooh."

"I love him!"

Then Michael hands her three cards. "I want you to open this one now. I want you to open this one after you take off, and I want you to save this one until you get home. Can you do that? In that order?"

She agrees, and opens the first card. On the tape, I can hear her gasp.

"What a cute card!"

"You like it?"

"I'm gonna cry!"

"You're gonna cry?"

"I aaam!"

Michael puts the other cards in her bag. "Are these going to make me cry even more?" Meredith asks.

"I don't know," Michael says.

They talk about the convention for a while, and then Meredith tells Michael how grateful she is for all he's done for her during the week.

"It's all been because I wanted to," Michael says. "Nothing's ever been work at all. I just wanted to—every minute of every day. Since the day I knew you were coming, I think that I got more and more excited as we got to that last week. And the minute I—I'll never forget the moment when I saw your face."

"I'm gonna cry," Meredith says again.

"I've never in my whole life ever been so affected, so quickly, so much. And to see all these other people notice that I notice just makes me know that much more how right I am—to see all these people just wanna touch you, wanna know you. I'm the lucky one. I am."

"No, *I'm* lucky. So many people at that convention would love to be with you."

"But none of them hold a candle to you."

They go on like this until the flight is about to leave. At the last minute, Michael kneels at Meredith's feet and hugs her with an intensity that's visible across the room.

Then Meredith walks down the ramp, a small wistful figure picking her way past the elbows of oblivious giants.

. . .

Walking back through the bustling airport, Michael starts to cry. "A lot of this is exhaustion," he says, wiping away the tears.

We pass two gates before he starts to worry. Meredith is going to Club Med in a week, he says. "Knowing her, there'll be some guy who lives in Connecticut who's a twenty-six-year-old investment banker and five eight . . ."

He talks about being lonely. He talks about going home to an empty house. "I wanted love for so long. Every year I say, 'I'm ready, I'm ready, but she has to show up,' and *she showed up*." He takes a deep breath. "At least the door is open," he says. "Nothing else matters."

Later that night, Michael seems slightly happier. Meredith called as soon as she got home. She misses him already. She's even talking about flying to L.A. to see him again. "I'm going to do everything I can to marry this girl," he says.

He leaves the next morning, skipping the closing-night ball. Without Meredith, what's the point?

. . .

Evelyn calls Dr. Carson with a list of questions, hers and Jocelyn's and David's too. First, are they going to operate from top to bottom?

"That will depend upon an MRI," Carson tells her.

"An MRI?"

"You need to have an MRI done," he tells her.

A few phone calls later, Evelyn discovers the MRI will cost $3,500, and probably neither her private insurance nor Australian Medicare will cover it—they won't cover any medical treatment in a foreign country unless someone's bleeding to death. Putting off a call to Australia, she rings the LPA office downstairs and explains the new crisis. Quickly a squad of helpers mobilizes: "Mac" McElwee of the Little People's Research Fund sets up a command post at the foot of Jocelyn's bed, Yvonne finds a room for them at the Baltimore Ronald McDonald House, Sharon arranges transportation and changes the airline tickets. When the Atlanta hospital balks at doing the rush MRI, an LPA member works the phone. "They're gonna have to fly out tonight because maybe Dr. Kopits can get them into an MRI study that he's doing in Baltimore. As soon as they find out, they're gonna call me through the hotel. Then we're gonna get them over to the airport."

When Evelyn finally calls David, she doesn't know how to tell him what's happened. Finally she just blurts out, "I need $3,500." Before David can come up with much of a response, she starts explaining about the MRI and the stress of it all and the hospital in Atlanta being ridiculous and what Carson said about losing the spine and she has *no control over anything anymore* and they are just going to *have* to let things play out. "If we go, I might have to go really quick and I won't have time to ring you, so I just want to warn you—if you ring and there's not an answer, I don't want you worrying about where we are."

"See if you can ring me before you go," David says.

. . .

Upstairs, Dr. Kopits works the phone. "There's a financial problem of some significance, and the little people have organized and helped with the plane tickets. I wondered if you could do a reduction in her case?"

A moment of Kopitsian silence follows—again I see him wait and wait and wait until the person he's talking to crosses the moral boundary from ordinary life to his world. Then he continues. "That would be great," he says.

While Kopits talks, Jocelyn walks around the room, even though her face is going pale and blotchy from the pain. She's getting more and more worried about the MRI. A kindly little woman named Mary Alice Johnston pats her on the arm. "Dr. Carson did the surgery on my daughter," she tells her. "You've got the best, kiddo."

Meanwhile, Evelyn is adding up the numbers and trying to figure out how she can pay for all this. When Kopits gets off the phone, he sees her distress and tries to soothe her. "If you wish, you could stay at our house on the hospital campus," he says. "The Pierre House, where the parents of our patients stay. It's very inexpensive, but good."

At the word "inexpensive," Evelyn gives him a look of such stricken gratitude that he smiles again. "It's a matter of having a little fund-raiser or cake sale," he adds.

Then he goes through the door to the adjoining room, where a dwarf is waiting for his examination. His nurse shakes her head after him. "He starts at nine and goes to midnight almost every night," she says. "He hasn't been outside for two or three days. Hasn't even been downstairs except for his breakfast."

Evelyn starts moving toward the door, fretting about all the things that need to be organized. But Jocelyn still doesn't want to get into her wheelchair. Finally Evelyn just points at it.

Jocelyn scowls and gets in.

Four

ONCE UPON A TIME THERE WAS A FAMOUS NAZI DOCTOR who stood at the train depot at Auschwitz deciding who would live and who would go to the gas chambers. A trained geneticist, he took advantage of his power over life and death to do experiments on various camp inmates, apparently in an effort to figure out how to make hair blond and eyes blue. But like so many people who dedicate themselves to perfection, Dr. Joseph Mengele seemed even more deeply obsessed by imperfection. "Of the millions who came to Auschwitz," writes a camp survivor named Sara Nomberg-Przytyk, "Mengele loved to single out those who had not been created 'in God's image.' I remember how he once brought a woman to our area who had two noses. Another time he brought a girl of about ten years of age who had the wool of a sheep on her head instead of hair. On another occasion, he brought a woman who had donkey ears."

Mengele had a particular interest in dwarfs, eventually building a collection of about forty. He was especially excited the day a whole dwarf family arrived on the train. The father was a circus performer, but the mother was a "tall, strong woman" who had given birth to six dwarf children, three boys and three girls. One of the sons had married a tall woman who gave birth to a normal child. Mengele was particularly fascinated by the sexual possibilities and sent the dwarf men off to the barracks so he could ask the tall women over and over for every intimate detail. How did they do it? What was it like? And was the dwarf really the father of the tall child? When they answered all his questions, he took the boy off to his operating room and experimented on him until he died.

That's where most accounts of this story end. But Nomberg-Przytyk's version is particularly merciless. As the boy was dying, she wrote, a prisoner named Bibi came into the clinic with some very droll news—apparently the patriarch of the dwarf clan made such a terrible nuisance of himself asking his fellow prisoners to help him get news of his family, they got annoyed with him and told him to take advantage of his small size and slip under the wires separating the men's area from the clinic. He was too small for the guards to see him, they said. As Bibi told this story she "was laughing uncontrollably, as if she had just heard an unbearably funny joke, and at the same time she was watching us. She was waiting for us to join in the laughter."

The punch line was one dead nuisance, the old dwarf-gladiator joke with an Auschwitz twist. Because the dwarf was small enough to slip under the wires but not nearly so small for the guards to overlook. And Nomberg-Przytyk brings it home: "We left the infirmary. From a distance you could see the dead midget. His fellow prisoners were as much to blame for his misfortune as the SS man who actually put the bullet into him. That was the tragedy of Auschwitz."

I won't read this story until much later, but somehow I already know it. And I don't think this is superstition or romanticism. As Fiedler puts it, speaking of the "priestly executioners who dispatched monstrous children at birth," this is one crime where we don't need to see evidence: "This history tells us, and *the response we feel in the presence of Freaks confirms.*"

My italics.

. . .

For the last two days I've been looking for Martha. Yesterday I stopped a tiny diastrophic woman and asked her if her name was Martha. She had the same short arms, the same robin's-breast chest. She wasn't wearing pearls, but I thought maybe she had taken them off. Sometimes the rule of genetics is so strong that dwarfs of the same type look startlingly similar, even to other dwarfs. One night I heard a pseudo-achondroplastic dwarf mistake another pseudo for yet another pseudo, asking him if he'd gone up to change his clothes. They laughed about it. So now I see another three-foot woman with a preening chest and stubby arms, and I'm not sure—is she the woman I asked yesterday, or is she the real Martha? Or some other three-foot woman altogether?

Only one way to find out.

"No, I'm not Martha," she says in an irritated way, as if she's talking to a particularly stupid person. "You asked me that yesterday."

The next time I see her, a couple of days later, she gives me a wary look—no, I'm still not Martha, you big idiot.

. . .

According to the social scientists who have studied them, young dwarfs tend to be shy and aloof and introverted. They maintain "firm boundaries between themselves and the outside world." When they draw, they tend to omit body parts like eyes and to use more shading than average-size children, which is considered a sign of anxiety or graphic sophistication or both. This puzzled the scientists who first discovered it but seems to suggests that for dwarfs—and perhaps for all of us—there is a powerful link between anxiety and eyesight.

Which brings us to the Marriott swimming pool on this sunny Wednesday afternoon. Andrea is wearing a backless white dress with blue flowers. We've worked out our little misunderstanding, and now she's telling me about safe sex. "There's concern that the teenagers are fornicating wildly," she says, her voice droll. In fact, she just came from Beth Tatman's sexuality workshop, where one of the topics under discussion was condom distribution. Most of the dwarfs are against it for the usual conservative reasons, the fear of encouraging sexual license and such. It's depressing how conservative dwarfs are, she says. Aside from that, a lot of the class was about basic stuff like, How do you masturbate when you've got short arms and they don't quite reach down there? Some of the issues were new to her, because she herself has done it only with average-size guys. She says this very casually. "I've gotten naked with little people, but I've never fucked one. I don't want to hurt any feelings by being too specific, but it did feel bizarre. It was fine while we were lying down—it was nice because our feet came to the same place. But when he got up and I looked at his body, that was kind of strange. It didn't look like the male naked bodies I had seen before. Before it was always, you know, the legs kept on going."

Trying to match her casual air, I ask for her position on the long arm issue.

"Oh, they're handy," she says.

Then she starts crying, actually sheds a few tears and wipes them away. She acts as if I didn't notice and so do I, erasing the wet eyes from our mental sketch of the moment. "I'm starting to get more attracted to little people," she says.

. . .

Tonight there are dwarf lovers everywhere, walking hand in hand and sitting on steps and dawdling by the pool. There's a pudgy dwarf boy in a backward cap and black clothes holding hands with a pudgy dwarf girl exactly the same height as he is. There's English Jason with that perfect little California blond who looks like a miniature Pamela Anderson—Dawn, I think her name is. There's a little man in a velvet vest and evening jacket escorting a small woman in a short cocktail dress. Before long the single people come down from their rooms and a brittle electric charge starts to build. The plan is to go to Buckhead again and get crazy, because tomorrow night is the ball and then the week is over. But everybody knows the truth, even if nobody wants to say it. If they haven't hooked up by now, their chances are slim. Which means another year of loneliness and the Internet. That's what you see in Kay's face as she sits on the telephone stools, head hanging and eyes teary.

And here comes Andrea, showered and changed and looking downright perky. She sits down next to me and we watch the parade and pretty soon I realize she's getting awfully touchy. Every other sentence, her little hand reaches forward to touch my arm, my hand, my knee. I am telling her about my romantic troubles in the past and I use the phrase "a guy like me who's so—"

"Gorgeous," she says.

I know I should make some pleasant remark about being flattered or at least give one of those insouciant European winks to acknowledge our mutual imprisonment in the human comedy—the gods play with us as wanton boys with flies, that sort of thing. Something grown-up and relaxed. I shouldn't just ignore it. But then I notice how her arms bulge around the wrists and elbows, that baby-fat effect, and my mind goes numb. I just can't think of the right thing to say.

. . .

By ten o'clock, only a few forlorn stragglers are left at the hotel. There's Kay, looking like someone ran over her cat. "I had three romantic prospects," she tells me. "Well, really two. One was this guy I met in a chatroom on the Internet. We got along really good on-line and agreed to meet on Saturday. But . . . he said hi, and I haven't seen him since."

A dwarf named Pam sits next to Kay, quietly fuming. "They come up to you, they look at you, and if you're not what they like, they walk away. They don't really care what's inside."

Pam's quite pretty herself, with short blond hair and punky mascara, nicely turned out in pearls and a lacy blouse and black jeans. She herself has hooked up with someone these last few days. At least that's how it looks so far, which makes it even more painful to think of Kay and all her friends who can't find anybody even though they're as nice and outgoing as a person can be. "Last night I went into the bathroom and there were seven girls crying in there," Pam says. "Because of these men."

The really sad thing is, they knew it would be this way. That's why they met at the airport, just so they walk in the door together, both of them so tense they felt like throwing up. "What makes me sick is that an achon or a pseudo can't look at a different dwarfism and go, you know, she's a really sweet girl, she's got a great personality—so what if we're different, I'll take her. But they don't care. They want somebody within their own dwarfism. You put your heart and soul into this week—"

"It's so important," Kay says.

"All year long you don't find anybody and you come here and you see somebody in the same boat you are and you put your heart out there and it gets stepped on. It does, it gets stepped on."

She's raising her voice now, attracting glances.

"And it breaks my heart, because we shouldn't be treating each other this way."

Of course, it's human nature. Pam knows that. She knows that everyone discriminates and knows how hard it is for the more extreme dwarfs, the ones like Martha, the ones with Morquio's syndrome or SED or the diastrophics stuck in their custom wheelchairs. "They're really resentful," Pam says. "They're looking at us, and we have . . ."

"Mobility," says Kay.

"And they're saying, 'Well, that's not fair.' "

And they know that this is part of the attraction of the Little World—the chance to discriminate against someone else for a change, to reject someone else, even to make fun of someone else. "When you're with average-size women, you're sort of . . . not *compared* to them. They're all grouped, and you're in your own class. And when we walk in here, we're all equal, so we're all judged."

But sometimes it's just too painful, and you start to wonder if it's worth it. "To be truthful," Kay says. "I'm looking at the possibility of an average relationship. Opening myself to that chance. Because of the men here."

. . .

An hour later, they're still at the table. "Midget," Pam says.

"Short," Kay says.

"Fat."

"Freaks."

"Children are the worst," Kay says.

"The worst for me is when a child goes, 'Mom, Dad, look!' "

"And they look."

"And the parents don't say anything. They look too."

"They look too."

"But that's not us," Pam says.

"That's not us," Kay says.

Everyone wants to be different except for the people who really are.

. . .

After midnight there's a constant stream of people moving through the lobby. There's Jason holding hands with a small brunette in a black shirt and jeans, the little Pamela Anderson evidently having moved on. Rudy gets back from Buckhead where they were dancing and dancing and the music was hot. He gives two passing women a bleary grin. "How's my two favorite girls?" There's a cute little girl wearing a T-shirt that brilliantly combines the evening's themes with a red stripe at breast-level and the legend: MUST BE THIS TALL TO RIDE. And there's Michael with news of Meredith—she called when she got home and said she missed him so much that she went straight to the one-hour photo store and the pictures came out phenomenally. "I mean, she was *raving* about these pictures." He looks around and sighs. "She lit this place up. I'm gonna do everything in my power to marry her."

At one, Jason walks by with the brunette and also the redhead he was ignoring at the gym. As the three of them go down the hall past the Coke machine, the girls cuddle in, snaking their arms around his waist. And Jason hugs them even closer, resting his elbows on their shoulders like a little drunken lord.

At two, a man named Anthony Soares tells me that he'd like to register his objection to the presence of Disney. "Here we are having an expo for an organization that's supposed to promote advancement for short-statured people, and to see a group from Disney, where they put people in rubber suits and put prosthetic ears on us . . . it sets us back. We're the only disabled group people have a license to laugh at."

Which is why he wants to change the LPA name. "It sounds childish," he says. "I stumble over the words 'Little People of America.' I'm very uncomfortable saying it to anybody outside of the organization unless they're a friend who knew me for a long time. It sounds childish. It doesn't sound serious. And we're not 'little people'—we're human beings encased in a short-statured body, or body that's different, or disproportionate."

Anthony is a senior art director of Griffin Bacall advertising, a subsidiary of the ad giant DDB Needham. He looks like a standard achondroplastic, a bit solid in the forehead but nothing more, and he says that once people get to know him, once they've worked with him and spent time with him, they don't think of him as "a short-statured person with dwarfism" but as Anthony or even Tony. "And they never think of me as a little person, at least I hope not. So if I say, 'Oh, I'm going to my Little People of America convention,' it sounds pretty funny. It sounds like I'm gonna be, you know, swinging from chandeliers and having a good time, when I'm really coming to a conference where people who are short statured learn about disability rights and career objectives and helping other people and—some of these people have alcohol problems and high rates of suicide, especially the more disabled ones like the Morquios, who are basically gonna die before they're thirty-five."

It's two-thirty in the morning, and Anthony is coming on a little strong. "Is that true?" I tease. "Or did you come to dance and look for girls?"

The question startles him. "No! I mean I—I've—well that's part of it. But that's part of meeting other people and building families. . . ."

"Building families?"

"It's not a bunch of horny-toad midgets," he snarls. "That's what some people may think. Like some people here may even tell you, 'Oh yeah, it's a great time and you get drunk, and meet a girl and have a good time and dress up like a clown and wear a bad T-shirt that says REAL LIVE DWARF on it.' But it's not about that. I come here because it's a week at ease, not having to deal with the gawks and the stares and the comments and the questions."

Anthony doesn't think much of the sports activities either. "Let's face it—nobody at this convention is going to win a medal in the Olympics."

"What about personal best?"

At this point a drunken dwarf with an Israeli accent approaches us. "Anthony, your room's on fire," he says. Hint, hint.

Anthony ignores him. "I agree with personal best. But I don't want people to have their personal best this week and then go home and collect disability or

Social Security and not be able to find a job in the career they went to college for because of a stereotype. This can't be a false sense of reality."

The drunk guy pulls over a dignified-looking dwarf woman whom he introduces with elaborate irony as "the first ordained short-statured woman Presbyterian pastor," having fun with my presumed taste for journalistic clichés. Then he turns to Anthony. "Would you join us for a moment of prayer?"

But Anthony won't take the hint this time either. "Who are you writing about?"

I mention Michael and Evelyn and Joc—

He rolls his eyes, totally disgusted now. "Michael Gilden—that Guido stud."

"Guido stud?"

"Did you see the way he was all over the girl with the cane? That's the sort of thing that could chase her right back out the door."

The Israeli dwarf nods his head. Even the first-ordained short-statured woman Presbyterian pastor rolls her eyes. "He's the kind of guy who's just after one woman after another," she says.

"And he's in that Radio City Music Hall show playing an elf," Anthony says. "*That's* real good for our image." He goes on, getting more and more angry. Why not write about dwarf doctors and dwarf lawyers? There's a girl here who is three feet tall and a first-year medical student at Johns Hopkins University. *That's* the kind of person you should write about, not some sensational crap about an elf-portraying womanizer.

Swaying in place, the Israeli dwarf catches Anthony's fire. "You want to know what it's like? Try walking for a week on your knees."

Anthony winces, as if his friend has gone too far. Softly, mostly to himself, he mutters, "No. That won't really do it."

. . .

As the clock crawls toward three, teenage girls keep raging through the lobby. "Did you find him?" one asks. "Oh, they're talking," her friend answers. At a telephone, a girl coos. "She wouldn't do anything with you because she knew how much I liked you."

At quarter to three, Michael's friend Gibson gets back from a long night of barhopping, arm in arm with a girl in a red-speckled jumper, his debonair poise unruffled as ever. Gibson says they cruised from bar to bar in a pack of fifty, taking over entire dance floors. The tall people didn't know what to think.

Anthony Soares overhears him and shoots a line across the lobby. "Yeah, a bunch of drunken horny midgets—that's the image we wanna portray."

Another snarl of teenage girls comes raging through the lobby, all worked up over something. Emily's boyfriend comes along in their wake and explains: a sixteen-year-old girl was having a romance with a boy, and she caught him with another girl. Then she told all her friends. Now they're looking for the faithless lover. "I'm sticking around to make sure nobody hits anybody," Chris says.

. . .

And here they come, dressed in gowns and suits, 744 dwarfs and their various tall friends. The women wear corsages at their wrists, and there are lots of tan lines and sunburns. The teenagers go wild with cameras and goofy poses, throwing their arms up in the air and mugging and slapping five and running up and down the carpet. Electric wheelchairs zip up and down the hall. Through the double doors comes the sound of a live band playing inside the ballroom—"shake it shake it shake it baby now, twist and shout." There's Andrea twisting away, and Dawn and Simon and Jason and even the girl who caught her boyfriend with someone else. Someone breaks out some funny hats, and there's another round of almost frantic posing as people write down phone numbers and promise to stay in touch. Gibson Reynolds says hi to a dazed-looking dwarf named Ethan. "How late did you stay up last night?" Ethan asks.

"Six o'clock."

"I was up till six o'clock too! Were we together?"

After dinner, Kay and Pam start planning an early escape. "Everyone feels depressed," Pam says. "If not right now, then often, and bad. They just pretend they don't. It's all a front."

An angry little man approaches me. "Can I say something into your tape recorder? I wanna organize a million-dwarf march on Washington so people in this country can wake the fuck up and realize that there's people who go through life day by day not being able to reach countertops, not being able to reach drinking fountains."

Someone introduces me to Rick Aines, the former editor of the LPA newsletter. Sitting on a little go-cart, he asks what I'm writing about and bristles when I get to Michael's name. "He's not your standard dwarf," Aines says. "He's got it made. If everyone was like him, there'd be no social problems. Sometimes it's even, why is he even here? Why isn't he dating average-size women?"

Then I see her, the real Martha, wearing the same pearls. Her head doesn't even reach my hip, and there's something painful about the way her hips jog from side to side. By now I know that means her spine is fused solid. Her little arms stick out from her body like—God forgive me—the arms of Mr. Potato Head. They are so short that even when she's standing with her chest pressed to the edge of the sink, she has to use hooks to turn the faucets.

She's happy to talk about Michael. Thrilled, in fact. "I hate that guy," she says. "First of all, he's one of these guys who thinks that he's all that, okay? I've been coming to LPA conventions for about eleven, twelve years. And I've never been one of the girls who's been, you know, absolutely gorgeous, with a voluptuous body and everything else like that. But I've had friends who do look like that, okay? And I've always noticed that guys like him, the decent-looking ones, the guys who think they're all that—although I don't think he's that great looking—will come up while I'm having a conversation with my girlfriends and *not even look at me.* And that's always irritated me. And the new age of computers has given me a new voice, and we were on the chatline at AOL one night and I basically chewed him out for ignoring me whenever I'm with a friend who looks great and I'm just little old three-foot-tall me and he doesn't even say 'hi.' And he's not the only one who does that. And it's always made me angry because I have just as much of a personality as these girls do."

Martha is wearing a little black ball gown and looks oddly glamorous. She has a cigarette in one hand and a champagne glass in the other. Her dark bangs are moist against her forehead.

"So that's what I said to him at the basketball game," she continues. "'It may ruin this romance for you, but you deserve it.' A lot of these guys are like that. They think when they get to LPA they can treat us like dirt, go after all the little pretty girls and leave me and the others sitting around. Just because I don't look like them."

By this time we've found a bench outside, but when I sit down I realize that for her the seat's at waist level. I offer her a lift but she refuses, standing there with her cigarette and champagne glass and telling me that she was born in the mountains of southwestern Virginia and her parents are both teachers, both average-sized, that she teaches theater in a Virginia middle school and she's had not one but two relationships with average-size guys. "I've gotten around, I'm not totally lifeless." She laughs. "Actually I've probably got more experience in relationships, *real* relationships, than a lot of those pretty girls."

I've begun to feel a little starstruck, like I'm talking to a celebrity. Her radical difference gives off that strong a charge. "You're not so bad looking," I say.

"Yeah, but I'm nowhere near them. Let's just be frank. I'm not gonna say that I'm ugly, but I'm different-looking. I don't have the body. I'm not tall like they are."

She laughs when I tell her about mistaking her for other diastrophics, that if it wasn't for the pearls I never would have recognized her. She says her mother gave her the pearls as a symbol of unconditional love and she never takes them off, not even in the shower.

Then she introduces me to a friend, a little redhead who has been standing nearby while we talked. Her name is Jennifer Arnold and she (Martha proudly says) is a first-year medical student at Johns Hopkins University. When I tell her everyone has been bragging about her, Jennifer smiles modestly. She talks about how great it was to finish her first year and how Dr. Kopits inspired her to become a doctor and every word makes Martha smile wider. "I felt a connection to her immediately," she says. "I felt it as soon as I met her."

Jennifer smiles. "I feel a connection to you too."

"She's just like me. We've only known each other for a couple of days but—"

"I feel like I've known her forever."

They giggle.

Jennifer is three feet two, two inches taller than Martha. Both of them have

SED and can't help feeling a little out of the LPA mainstream, Jennifer says. "The achons—there's just something about them."

"There *is* something about them. I don't know what it is."

"I was trying to explain to my mother on the phone today. It's kind of like the population at home where you have the white majority—it seems like you're the minority."

"I don't want to insult people," Martha says, "but so many of these achon girls—they bond with each other. They don't try to make friends with us."

"I'm proud. If they don't want to be with me . . ."

"You should be accepted no matter what you look like or what you are," Martha says.

Jennifer nods. "That's why it was so upsetting when I didn't fit in at first. I thought when I got here I would meet new people and it would be a happy place. And it's not."

. . .

By one-thirty it's over. The band packs up, the ballroom starts to clear. And Andrea finds me. I tell her about Kay and Pam and Martha, and she tells me to lighten up. "I think something sad in you responds to these people," she says. She learned a long time ago that people who are different remind "normal" people of their own sense of inner difference, and sometimes they can handle it and sometimes they can't. That's what I have to understand. I'm not sad for them, I'm sad for me. And she has a point and it's something I'm going to think about later, but right now I'm too busy noticing the way she keeps touching me—patting my arm and talking some more and patting my arm again. Then she starts tickling me, her stubby little fingers digging at my rib cage. I don't know what to do so I just squirm and let her tickle.

. . .

In the lobby the next morning, people slump in armchairs like travelers marooned in an airport. With each departure, the energy drains out the door.

But there's Dr. Kopits, smiling and laughing, bending down to gather his patients into his arms and beaming with visible delight at each person he hugs. There's nothing dutiful about this. His love is a reminder that with each twinge of discomfort or even revulsion these people inspire—because they do look wrong, there's no getting around that—there is also a corresponding exhilaration, a liberating delight that comes from encountering intelligent humanity in such inappropriate packages.

Heading for the exit, feeling the same disorienting mixture of shame and exhilaration I've felt all week, I run smack into the tricycle baby and his mother. I've been trying to give them a wide berth, but here they are standing at the video games, practically blocking the hall. The mother holds out her hand, smiling furiously and even blushing a little. She says her name is Angela McTate and she's sorry for getting so upset before. "I hope I didn't scare you."

"Not at all," I say, trying to slip by. But she stops me because she's got something she needs to say. The thing is, she knows how shocking it is to see a one-month-old baby talking and riding a tricycle. But Joshua is really four and hates the way people stare. This is the first time in his life that he's been willing to ride his tricycle in public. He just wants to be treated like anybody else. That's why she lost it when she saw me talking into my tape recorder. It wasn't my fault. My reaction was perfectly natural. And there's more to the story: she had another dwarf baby before Josh who lived only ten hours, so when it was Joshua's time and the ultrasound showed how tiny he was, the doctors told her flat-out that he would not "live to be born." And if by some miracle he did, they wouldn't give him emergency medical attention to prolong what was obviously another doomed life. Well, growing up in North Carolina, daughter of a carpenter, she was pretty much intimidated by the medical profession, but she knew enough to know *that* was wrong. So she moved to a different hospital and Joshua turned out just fine—so healthy they sent him home after just four days. Since then it's been quite an odyssey. They go to the hospital so often the nurses joke about giving her the employee discount. And when he was thirteen months old, they made the trip to Johns Hopkins and met with Dr. McKusick, the famous geneticist who diagnosed so many of the more exotic forms of dwarfism, and even he couldn't come up with a definitive diagnosis of Josh. They can't even tell her how much he's going to grow.

At one point it got so bad, she even began to fight with God. That was a scary thing. "I was a very, very, *very* strict Pentecost, from what I call the old school. I was so fearful of God—if I even *thought* of getting angry with him—I felt like he would strike me. But after my daughter had died and then we had Joshua, I finally got mad enough to talk to him. I said, 'You made me, you gave me everything I have, and if you didn't want me to be angry or show anger, you would not have given me that outlet.' And so all of a sudden, I went outside and I screamed and I bawled and screamed until I just could not scream anymore. And he didn't strike me. And I understood then that he was sad because this had to happen. So I don't think it's a punishment anymore."

This year she quit her job as an accountant and started taking premed courses so she can make a contribution—not just for Josh but for all kids who are different. "They want to be treated like people," she says. "That's what they are, no matter what they look like."

Suddenly Angela looks alarmed, her eyes sweeping the hall. "I, I, I can't see Josh."

Then she spots him and relaxes, wiping away a tear. "So yes, I'm very overprotective. I lost one child already. I don't think I could make it if I lost another one of my children."

As we speak, Josh gets upset and starts squealing at his twelve-year-old brother, who puts him down on the ground and turns back to his game. Josh waddles over to his mother. "Bubba won't let me play," he says. It's a pouty child's voice and it's truly eerie, coming out of this baby.

Angela points him back to the video alcove. "You go in and tell him to pick you up and let you look at the screen," she commands. A minute later he's sitting up by the controls, staring happily into the screen.

So many good things have happened at this convention, Angela says. He hated cameras because the doctors used them so much to document him, but today he actually *asked* to have his picture taken. That's the kind of thing money can't buy. "I'm hoping he takes this home with him, I really am."

Then Bubba gets tired of holding his brother and puts him down. Joshua comes out of the video room with an intensely frustrated expression, plopping himself down on the floor. Angela gives a sad sigh. Then a little girl rides up on a scooter and says, "Hey Josh, what's up?" He just lies there with his cheek against the carpet, not answering, but the little girl gets down on the floor and bends her head close. "What, they won't let you play? I'll go tell 'em. Come on, Josh." Taking him by the hand, she leads him back into the video alcove room.

Angela beams. "See, he's got a little person friend."

She won't let me get away without a couple of fierce hugs.

Five

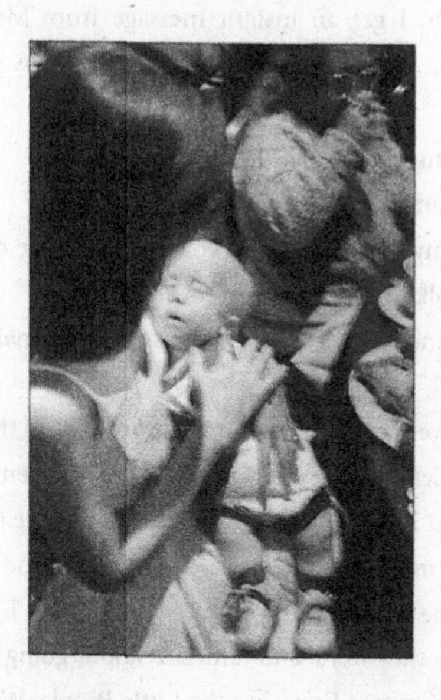

BACK HOME, MY OLD ROUTINES ARE WAITING: work, kids, tending my sick mother. And what a relief it is plunging back into "normal" life.

Then the first e-mail arrives. "Yes, we are home," Evelyn writes, cheerily narrating a saga of canceled tickets and battles with airline officials and a last-minute stop in Fiji due to headwinds and also Baltimore and the horrible MRI:

> *Jocelyn was okay for about two minutes, then she became hysterical and screamed to get out. The look of surprise on their faces! But they were so patient and supportive, nothing like in Australia where Jocelyn was nothing but an inconvenience. Then I remembered when I was in labor, how vital it was to hear David say, you're halfway. I knew that Jocelyn would be able to control her mind and the pain if she had a goal to work toward. It worked. I focused and hung on. When she came out of the tunnel after*

each set, she was crying and pale and after the last set she fainted. But she did it!

A few days later, I get an instant message from Meredith: "I'm back! Thought about you every day, ha ha. I'm going to LA on Aug. 6 to meet Michael."

I write back. "Can I come?"

"No you can't come!!!"

"BTW, how many hours a day do you spend on the computer? Do your parents know you talk to strange reporters on this thing?"

"I spend little time on the computer—just to check mail and have cybersex with dirty old men."

It's cheering, a welcome distraction from my trips to the Memorial Sloan-Kettering Cancer Center and the sight of bald children and men without noses. I imagine that I'm learning something about grace under pressure, and feel the urge to slip into the Little World again, to feel the buzz and swirl and sheer emotion of the convention. I think of the people I met and how open and honest and raw they were, and fantasize about going back next year and becoming, like Dr. Kopits, a Friend to the Little People. When I try to explain to my friends and coworkers why it was all so disorienting and emotional, I can't quite get it right and just blather on about how weird it is to be the only tall person in a room full of dwarfs. And when I see that people don't get it and aren't even all that interested, I feel a little smug—I know something they don't know.

On the phone, Meredith tells me she started missing Michael right after the convention, just as he predicted. They've been talking almost every day. "I just got a dozen roses," she says. "Eleven pink with one red."

Michael's just as happy. "Sometimes I pinch myself," he tells me one night, talking on the phone from California. "This is a girl who one minute sounds like a doctor of psychology and the next a six-year-old girl asking for chocolate milk. She's more extraordinary every time I talk to her."

Gibson Reynolds gets in touch too, asking me to read a few pages of the autobiography he's been writing and mentioning that he's going to be in my area. A week later we meet at the diner in my little town. Once again, he's about as cool as a guy can be, cracking jokes and deflecting my attempts at sympathy. A lot depends on your family, he says, and his parents were the kind who always said you can do whatever you want if you try hard enough. So yeah, he spent a lot of time in the hospital, but that was mostly because of his stupid habit of skiing like a madman instead of like a reasonable person. And yeah, it can be a drag to be little, but a lot of people are more sensitive and decent about it than you'd expect, like the folks he wrote about in his autobiography who waited so patiently as he went through the drive-in bank window on foot because he couldn't reach it from his car. They even turned on their headlights to make it easier for him. And the thing is, he thought they were being impatient. It took a moment to realize they were actually trying to be helpful. Which just goes to show you—paranoia is half the problem.

But once again things get complicated, this time because Gibson seems so relaxed and uncomplicated that I feel implicated every time an average-size person—one of my neighbors—stares at him. He's trying so hard to be normal. Can't we just let him? And that night on the phone, when I tell Meredith about the lunch, she's way too pleased. She says, "That's so *cool*," and another spear of guilt shoots through me. What's so cool about it? That a gloriously tall person like me was willing to have lunch with a pitiful little stub like Gibson? No! No! No!

Andrea gets in touch too, sending a friendly hello via e-mail. Pleased to hear from her, I send back a brief note that ends with a question about how it is to be back in the average-size world. It seems like a natural thing to ask.

"Hate to be a bore about this," she responds, "but regarding your question about life in the average-size world, are you asking as a friend or as a reporter?" She says she has an "impulse to share things" with me but can't help worrying that it will bore me or that it will end up in something I write, which is the same feeling she had at the convention when I asked all those questions about her father. Was I just looking for material? Was she burdening me with too much detail? And anyway, the answer to my question is complicated because moving between the big and the little worlds is no big deal these days but it used to be a really big deal, full of feelings of anger and hurt and . . . but she doesn't want to talk about it until I clarify the friend-or-reporter issue.

Then I get another note from Evelyn. "I need to 'talk' to a friend—OK? I am going to transcribe a letter that Dr. Kopits faxed to me this morning:

This is to certify that Jocelyn Powell is a sixteen-year-old achondroplastic girl who has severe neurological complications of her achondroplastic condition which threaten her with quadriplegia. Specifically, she has a very severe compression of her cervical spinal cord at the junction between skull and neck, compression as well as instability at the thoracolumbar junction in the middle of her back and severe compression of her lumbosacral spine nerve roots to the legs, in the lower back. She is in need of specialized neurosurgery and orthopedic surgery to the back and neck to be performed as soon as possible in order to prevent full paralysis. The special expertise to perform these procedures exist here in Baltimore. We have advised that these surgeries take place as soon as possible.

"When I read this," she continues, "I was in shock. I mean shock. I didn't think that anything could have elicited that emotional reaction from me. Up 'til now I thought the terrible end would be paraplegia. I have taken a lot of deep breaths today."

This begins a flood of e-mails from Australia. Evelyn writes to me every day, sometimes three and four times a day. One of her constant themes is how much she misses the United States and "all the wonderful support" she found, and almost immediately she starts mulling over the possibility of moving to Atlanta someday—Jocelyn's idea, but one she enthusiastically seconds. People told her this is what would happen to them, this urge to keep the magic of the convention going long distance. And it did. And she's also having a hard time explaining to people exactly what was so special about the convention. Her first Sunday back, her brother brought his family for a visit and asked her all about America. What was Atlanta like? Were the streets dangerous? Did they go to the Hard Rock Café? He didn't seem the slightest bit interested in what she was trying to tell him and didn't want to hear anything about Jocelyn's problems, and finally Evelyn got so angry she didn't even try to hide it—*who gives a flying sod about the Hard Rock Café?* The next morning she walked into her boss's office and burst into tears, saying that she tried to give Jocelyn the best care, really she did, and now look where they were—facing paraplegia and worse!

But she rallies. With the help of her boss and coworkers, she starts organizing and hustling, and within days the Jocelyn Powell Appeal Fund is off and running. At Jocelyn's school, where they've already installed an elevator and curb cuts and turned all the doors upside down so Jocelyn can reach the doorknobs, they immediately agree to help. As Christians, they believe that helping Jocelyn has been good not just for her but for them as well. After Evelyn leaves that day, she gets her first donation, $200 sent anonymously to her house. Buoyed, she fills her house with daffodils, the Australian symbol for hope, and books return tickets to Baltimore—for September 1, six weeks away. But the stress is almost overwhelming. When her computer crashes as she's filling out fund-raising forms, David tries to "take over" and they get in a snarling fight. Another day, she's late to an interview with a newspaper reporter and smashes her car into a pile of landscaping bark. Another day, Jocelyn is looking particularly pale and tired and the chiropractor tells them that her spinal cord seems to be getting tighter and Evelyn goes into overdrive—she needs to find someone to write a good flyer! She has to talk someone into making thousands of copies for free! How is she going to get it all done!

There are also great highs. One day a coworker comes in with a suggestion: if they hold a statewide "mufti day"—an Australian fund-raising method where they let schoolkids make a small donation to wear casual clothing instead of school uniforms—and every child brings in just one dollar, that would be $600,000 right there! So they fax out a flyer, and soon twelve schools fax back to say yes. Then ten more. In short order, the Appeal Fund mailbox is packed full every single day. Some of the letters they get are so touching that Evelyn and her coworkers write out the receipts in tears. A widower sends $5 from his tiny pension along with a note remembering his own handicapped daughter; an old woman writes to say how happy her life has been and that's why she's so grateful to be able to share that bounty with poor Jocelyn. And then Evelyn notices the fax machine ringing and it's another school agreeing to hold a mufti day—and she bursts into tears.

During these weeks my own relationship with Evelyn starts to change, mostly because she keeps pushing me to drop my professional detachment and talk about my own life. Mixing lecture with confession, she tells me that before she came to America she always tried to deal with everything herself and "just absorbed" all the emotions that got stirred up—maybe it was something from childhood, maybe something to do with the Australian frontier character, but she figured that everyone else had problems of their own and it wasn't fair to burden them with hers. It took Atlanta to teach her how important it is to share the pain and the fear and the ups and downs with other people, people she could trust. "I don't mean to preach but we have that relationship—don't we? One where our partners have nothing to worry about, but probably wouldn't understand. This has been on my mind for the good part of the day."

So I start to tell her some of what's going on in my life, like the stress test my mother took before starting her radiation treatments. It was so hard she broke down crying afterward, and then cried some more because she was ashamed of crying. And the fight with my dad over a piece of pie. He's always so self-denying—no pie for me, pie is for the foolish young, I'm going to die soon. EAT THE PIECE OF PIE! And his eye is all screwed up with glaucoma but he doesn't want to get it checked out—one eye is enough for him. Argh!

Soon afterward, I pick up the phone and hear Evelyn's voice. She's calling all the way from Australia. No crisis, she says, she just wanted to hear my voice. But she seems giddy and brittle and for the first time, I think maybe something is really wrong. What's going on down there?

Students of disability agree that "normal" people are far more likely to open up to people with disabilities when they have experienced something painful or debilitating themselves, or when they love someone who has. No wonder, then, that Evelyn and I turn out to be each other's perfect confidant. With her daughter and my parents both so sick, we're in very similar positions. And we're the same age, both juggling jobs and children. With each other, we can talk about the queasy feelings only another caregiver can appreciate, the anger and frustration and hunger for escape and the pressure to put on a brave front and the urge to blurt out the ugly truth and the corresponding urge to tell bland prophylactic lies—and how we can't help resenting the people we've misled for their annoying innocence. And how the doctors dole out bits of truth and how you want to take over everything and just run the world and how you can't seem to shut off the droning madman in your skull—my we're looking gaunt in that passing window, clearly someone probing life's darkest corners. And how sometimes you don't feel anything at all and start to think maybe you're one of those people who are just faking human emotions they've learned from movies—fond of children, loving to parents, pious at church—until the moment when you're sitting at a stoplight or waiting in a hospital hallway or writing an e-mail to some woman in Australia and the feelings come rushing in. . . .

As I open up, Evelyn bursts loose with more and better. Like the day her mother makes a dig about her alleged spendthrift ways. "Explain to me how strangers, people we never met, are working their butts off to raise money for Joc and this woman can be the way she is?" Or the day her best friend asks her if she'd be willing to postpone that September 1 departure date, and the suggestion alone makes Evelyn so jumpy she barks: *No way!* "The only thing keeping me going is the thought of getting on that plane," she writes. And then there's her husband, who *insists* on coming along and has been going on and on about how distant and remote she's been since she got back. Can't he see she's got more important things to worry about? She doesn't have time to deal with his *ego.* "I admit I am keeping all emotion at arm's length at the moment because it is just that overwhelming and I need to stay in control," she confides. "But I am starting to teater. Help!"

I write back, thrilled to be caught up in Evelyn's problems instead of mine. "You're entitled to 'teater,' whatever that is. You don't have to be Superwoman *all* the time."

She responds instantly: "It looks as though 'teater' is an original word in the special language of Evelyn!! I can't find it anywhere in the dictionary and yet it is one I use often. Meaning 'to be unsteady, trying not to collapse.' " Signing off, she adds, "Do you know you are the only person in the world who is allowed to call me 'Ev'?"

Soon this e-mail relationship has become part of my daily routine. I pitch in whenever I can, writing a press release and a fund-raising letter, editing some of her own work, advising her on how to deal with the local media—stick to one or two sound bites, restrain your impulse to trash the Australian doctors, that sort of thing. In return, I get stories from across the sea, the vast distance they travel somehow part of the charm. Like the night two of Jocelyn's teachers make good on their pledge to shave their beards the day the kids raise $1,000. And what long beards they were! Uncut for decades! Reading the story the next morning, I chuckle and feel warm and a little special too, pleased again to be part of something the "normals" don't know. Or the day she goes to work feeling awful, all jellylegs and cartwheel stomach, and no sooner does she sit down than she gets a call from a bigwig at the Department of Education who saw the appeal fax and wants to help—in fact, he's already called a friend at Australia's leading television newsmagazine, *Today Tonight.* Or the Famous Bike Ride of Bob the Blind Man, which comes about when a man named Steve calls Evelyn out of the blue and says he wants to raise money by riding a two-man bicycle with his blind friend. A blind man on a bicycle? She doesn't know what to think of that, but it's a new world and what can she say but sure, thanks, of course, launching herself on another lesson in life's strange riches:

> *To see them ride is quite a sight. There is a really steep hill near where they live and as they are riding down it Steve will yell out to Bob to stop peddling* [sic]*—but with the wind rushing past his hearing aids all Bob hears is whistling—so he keeps peddling hard—at the bottom of the hill an intersection—Steve is having heart attacks they won't stop in time because Bob is still peddling!! And then I look at them and think wow—these are strangers.*

During this time, my relationship with Andrea is also getting more involved, not entirely in a good way. After her last note, I promised again not to quote her without permission and tried to clarify the friend-or-reporter issue as honestly as I could, admitting that I was curious as a reporter but also tended to gravitate toward people I found genuinely interesting. That doesn't solve the problem. In her next note, she says that she's come to realize that her fear of talking to me is actually "about more than what I originally said." Her real fear is that when I finish writing whatever I end up writing, I'll stop being interested. Again, I try to convince her that the writer in me is no different from the person, that I hung out with her at the convention because she was funny and smart and that I've become friends with a number of people I've written about over the years. All of which is true. And she responds with the same moving honesty, admitting that my curiosity challenges her boundaries and that it's "wonderful and scary" and she just doesn't know how to handle it. She finally answers my question about coming back from the LPA convention to the normal world—when she first started going to the conventions, she says, being around so many little people gave her a feeling of comfort she'd never before experienced, a rare taste of freedom from scrutiny, the bliss of being anonymous in a crowd. But afterward she "had this experience of coming home from the convention and being extremely angry (I suppose at average-size people)," a feeling so intense it scared her. With rage like that, how was she going to live in the average-size world? How was she going to stay close to her average-size friends? And she *had* to live in the average world. It's great material, and I'm pretty sure she knows it. She's dangling this delicious insight in front of the hungry mediabeast to see what it will do.

Clearly there's only one decent response. So for the next few weeks I stick to everyday stuff like her search for a new car and problems at work, sketches of the day and stray thoughts, and little by little she starts telling me about the deep and painful rift with her father, who was so remote and full of judgment, who was never physical and never made her feel loved. I respond with stories about my father and childhood and the funny thing is, the more it becomes a real friendship, the more we return to those prickly issues of height. We can't avoid them. And somehow they take us back to a deeper level of the more ordinary stuff. One day she tells me that her father is sick and may be dying and asks if I think she should visit him, and I realize that she's asking me because I've become some kind of bridge across the boundaries between short and tall—a bridge back to her father and even to herself. It's so deeply satisfying, I don't have the sense to be apprehensive.

And Evelyn is losing her voice. She's having dizzy spells. Then she comes down with the flu and still shows up "half-comatose" to watch a thousand kids spell out "Send Jocelyn to the U.S." in coins on the school plaza. In contrast to Andrea, she tells me all this so willingly, so eagerly, so completely without suspicion. Gratefully, I soak up the details: Collected, the coins add up to $1,343.60, two newspapers show up to take pictures, then it's on to the football fields where Jocelyn and Evelyn and dozens of helpers pass out flyers and man the contribution buckets. This time they raise $2,500. But even as the coins clink in the buckets, Evelyn feels a building anger. All these men bashing into each other—what's the point? Why would they bother with something so stupid when Jocelyn is suffering?

It's that outsider suspicion again, aimed at the footballers instead of me. Odd to see it from the other side. To distract herself from her anger, Evelyn asks Jocelyn about her plans for finishing school. Will she go back to William Carey or try to test out? And Jocelyn says that as far as she's concerned, she's finishing high school in America—and the rest of her life too. Evelyn's stunned. As the big dumb men crash into each other inches away, Jocelyn says that she's found where she belongs—found her people. The connection she made in Atlanta is real and can't be broken.

Divisions, alliances, differences: it's as if illness has marooned them on a living sea, and the bodies crashing like waves around them are so close and yet so alien. It's disturbing to be so remote in the midst of life. And Jocelyn's legs hurt and her back is going numb and the game goes on and on and it's just one more thing to get through. That night, in the refuge of her bed, Evelyn broods. Will Jocelyn really stay in America? What about the family? Where will this end?

Two nights later, the *Today Tonight* crew shows up to film another fundraising evening at Jocelyn's school. Her classmates do a song-and-dance routine and the principal unveils a huge posterboard check for $20,000 and it's all so moving that David breaks down in tears. The stoic, reserved engineer himself! While the camera was rolling! "Poor old David. The only other time I have seen him cry was when I first told him about Jocelyn's diagnosis. It shows how much he has absorbed and how stressed out he is."

As for her, she's determined to control herself. "I'm keeping emotion at arm's length at the moment. But the *Today Tonight* crew is coming to the airport, and I'm not sure I'll be able to avoid teatering then!"

. . .

During these months, Meredith and Michael keep working at their long-distance romance. There seem to be a few rocky patches, but they don't want to talk about them. Meredith tells me about dancing with Michael that first night at the convention, and how he sang to her on the phone. "Oh my God, he sounds *exactly* like Steve Perry." She talks about her summer job at the United Cerebral Palsy Foundation and her plans to go for a Ph.D. in developmental disabilities because of her brother's epilepsy and learning problems. She's cheerful and upbeat.

Michael's more distant. On the phone from Los Angeles, he says he's playing a role in the next Billy Crystal movie, *My Giant*, and just did a Taco Bell commercial with three midgets. "I used the expression 'little people,' and they said 'Fuck that, we're midgets.' "

At times he sounds depressed. When I ask him about his childhood, he tells me that he didn't know he was little until he was almost ten and saw some home movies and it hit him all of a sudden: his little brother was so much bigger than he was! How did that happen? What did it mean? And other kids noticed too because that was when they started to make fun of him. "Then you're eleven or twelve and you're starting to like girls, and you realize it's not going to happen."

Sometimes he gets very sour. When I ask him about his decision to play an elf in the Radio City Music Hall show, he snaps at me. "It's not like being a circus clown, you know. It's a huge historic show."

"I don't get you, Michael. You're so sensitive, and the opinion of other dwarfs means so much to you, but then you go and do the Howard Stern show."

He bristles. "What I did wasn't bad," he says. "It was me and Kristopher and a guy named Marty, and at the time Marty and I were dating average-size girls, and Marty just decided to say some things that were questionable. He doesn't like little women, and he made it clear. I said I adore the opportunity to date little women, I just happened to have been dating a tall girl at the moment, but it was guilt by association. Because we were all on the show together, they assumed we all felt that way."

Then he breaks off. "You know, I don't think that highly of myself," he says. "I've been in a major depression."

When he's not talking about his anxieties, Michael leans to the subject of women. When he mentions for the fifth time how beautiful his ex-wife was, I ask him if beauty is important to him. Again, I can almost hear his hackles rising. "I was talking to Meredith sight unseen for five months—it was a complete shock she was so attractive."

"I thought she sent you pictures."

"The pictures weren't that accurate," he snaps. "I don't think I'm superficial at all. Meredith walks with a cane. A lot of guys might have been turned off by that. It's in her eyes—it has nothing to do with the physical beauty."

. . .

After one of these prickly conversations, I dig out my old grad school copy of *Freaks*. It's right there on the bookshelf behind my desk, untouched for fifteen years. Flipping through it, I stop again on the excerpt from the *Member of the Wedding* when Frankie Addams visits a circus sideshow. Lonely, sexually confused, growing so fast she's afraid she'll end up nine feet tall, she stops in front of a hermaphrodite dressed half as a man and half as a woman, and realizes that she's afraid. The freaks "had looked at her in a secret way and tried to connect their eyes to hers, as though to say: *we know you. We are you!*" This time, even though I know it's the character talking and not necessarily the author, the passage strikes me as false and literary. I've seen nothing like this in the eyes of the dwarfs I've met. More the opposite, open faces with no secrets, as if they were all trying to mimic the breezy well-being of suburbanites in a Dockers ad. And instead of an insinuating *we know you*, the almost imperial disregard of people who have learned to cut their way through curious crowds. I know that McCullers often used freaks and dwarfs and minorities as symbols of the worldview she called "queer." In *The Ballad of the Sad Cafe*, revealingly, a dwarf is the only man the defiantly hyper-masculine heroine can allow herself to love. And I know that Fiedler was inspired by the social upheavals of the 1960s and the hippy transvaluation of the epithet "freak" into a badge of pride, that he was conscious of being part of a literary and intellectual world that seemed to be going rapidly out of date. He couldn't help wondering if this brave new world of liberated women and guilt-free sex wasn't going to take the tortured romantic grandeur out of life. So he asked, almost wistfully, if his generation wouldn't be among the last "whose imaginations would be shaped by a live confrontation with the nightmare distortions of the human body." Because the sideshow to him was more than a mere amusement, more than an opportunity for personal psychological insight. It was a hint of the sacred. The sight of "human monsters" threatens (he said) the boundaries we draw around ourselves and reminds us of "the thrill our forebears felt in the presence of an equivocal and sacred unity we have since learned to secularize and divide." And again this seems to me an uncomfortable mixture of wisdom and a shocking lack of human tact—until I turn to more contemporary writers and see that his predications have almost completely come true. Almost all the current fiction with dwarf characters is solidly in the "big hearts in little bodies" mode. The wistful hero of *The Dork of Cork* spends much of the book staring through a telescope at the unreachable beauty of the stars; the plucky little heroine of Armistead Maupin's *Maybe the Moon* wisecracks her way to love and stardom; John Irving builds *A Prayer for Owen Meany* to the little hero's redeeming act of bravery, and it's all so sentimental and uplifting you can almost see the sampler stitching. Which is not to say that these books are completely lacking in insight—one of the things about dwarfs is that it's almost impossible to think about them without getting hit by some flying chunk of wisdom. Like this passage from *Maybe the Moon*:

> *When you're my size and not being tormented by elevator buttons, water fountains, and ATMs, you spend your life accommodating the sensibilities of "normal" people. You learn to bury your own feelings and honor theirs in the hope that they'll meet you halfway. It becomes your job, and yours alone, to explain, to ignore, to forgive—over and over again. There's no way you can get around this. You do it if you want to have a life and not be corroded by your own anger. You do it if you want to belong to the human race.*

Even Ursula Heigi's tough-minded and lyrical *Stones from the River*, probably the best dwarf novel ever written, hits square on almost every politically and therapeutically correct point (abuse of women, fear of the Other, the subtle gleam of inner beauty, the ignorant normals versus the courageous outsider). Ultimately, the secret purpose of these books seems to be aimed at taking the sting out of dwarfism, and the irony is that this is something dwarfs themselves seem to resent—they often complain about being compared to elves and fairies, for example, but I have never heard a single one complain about trolls. It's as if they sense that the sweetness of elves is a covert way to defang them, to evade the questions they ask about justice and the limits of our decency.

But what if they reject human justice? And even humanity itself? This is the fear the darker books address. In a recent novel called *Geek Love*, a mother who works in a circus takes poisons to stimulate the birth of deformed children who grow up reeking hatred for the "norms" who gawk at them. "We are masterpieces," sneers the narrator, an albino hunchback dwarf. "The only way you people can tell each other apart is by your clothes." In *Mendel's Dwarf*, a dwarf geneticist meditates on chance and heredity with the most bitter intelligence, ultimately deciding to abort his own child. Although sometimes these books seem almost programmatically dark, the literature of graduate students who have read too many *ecrits* about the Marquis de Sade, they get closer to what feels like the truth. And most of these books seem downright responsible next to the high-lit traditions of yesteryear, where dwarfs almost universally seethe with resentment and spite. The model here is *Rumplestiltskin*, where the bitter dwarf demands a bribe of children—note the genetic fear at the basis of the myth. In Edgar Allan Poe's *Hop Frog*, a king's pet dwarf convinces the king and all his ministers to dress up as apes and burns them to death. In *The Tin Drum*, Günter Grass imagined dwarfism as a willful act by a boy who refuses to grow into an evil world and spends his time banging out his disgust on his magic drum. Over and over, dwarfs embody the great refusal. But none come darker than Pär Lagerkvist's *The Dwarf*, a bracingly insensitive novel written in the last years of World War II. Romping through all the old myths, Lagerkvist tells us that dwarf children never play and that dwarfs are "ancient wizened guests on a visit which has lasted thousands of years," clearly nonsense from a literal point of view (and no doubt infuriating to actual dwarfs), but still eloquently resonant of our inner dwarf, the part of us stunted by hate and self-loathing. Because Lagerkvist's great theme is the bottomless anger we all feel, the frustration at our helplessness and the vicious stupidity of the big world around us, and whether or not it's nice to say so, something deep in our archetypal imaginations seems unable to resist projecting that anger into a little homunculus demon doll. His dwarf sneers:

> *I have noticed that sometimes I frighten people. What they really fear is themselves. They think it is I who scare them, but it is the dwarf within them, the ape-faced manlike being who sticks up its head from the depths of their souls. They are afraid because they do not know that they have another being inside of them. They are scared when anything rises to the surface, from their inside, out of some of the cesspools of their souls, something they do not recognize and which is not a part of their real life.*

When people try to ignore that inner dwarf they "go about tall and unconcerned, with their smooth faces," reason enough to hate them. But sooner or later, the hidden deformity pushes back to the surface and they become confused and lost. And the dwarf hates them even more. "I live only my dwarf life. I never go around tall and smooth-featured. I am always myself. I live *one* life alone." He is the absolute man, the ultimate undivided soul, so completely known to himself that he despises the mysterious and the unknown as unworthy. "There is nothing 'different' about me," he sneers—and in the end, contrives a mass murder even more spectacular than the one in *Hop Frog*.

Writing under the specter of Hitler, Lagerkvist made his dwarf into a monster of *ressentiment*. Thirty years later, Fiedler did the exact opposite, calling dwarfs "the Jews of the freaks" and pointing out a surprising number of correspondences between little people and his people: like the Jews in Europe, they found patrons among the nobility; like Jews in America, they found a haven in show business; like Jews everywhere, they are of outsiders "the most favored, the most successful, the most conspicuous" and also "the most feared and reviled, not only in gossip and the popular press but in enduring works of art, the Great Books and Great Paintings of the West." In conclusion, he gives the argument a lovely turn that mixes an ancient fatalism with a transvaluating pride: "They have been, in short, a Chosen people, which is to say, a people with no choice."

We imagine the dwarfs we need.

Six

HOSPITALS DO THINGS TO YOU. Say for example your dad admits one night that he hasn't peed for a week and he's about to explode and you rush him to the emergency room and they stick him on a gurney and take some tests and then forget him for a few hours and you notice ridiculous things like how slender and girlish his legs are and then one doctor comes in and says he has heart failure and another doctor says no it's something else and then they forget you for another hour and your dad says this is torture just let me go home and you go off to find a doctor and when you come back you find the old man crawling off the gurney and the nurse says she can't send him up to a room or give him any drugs until the doctor finishes writing his report and the doctor can't be found and finally you lose it—*He's eighty-three years old and exhausted and why can't the fucking doctor write the fucking report a little fucking faster!*

Say these things happen to you. A therapist might diagnose what you are going through as some kind of hospital mania. He might compare it to being at war because there's the same weird boredom of sitting around and sitting around and sitting around, idly playing cards with death and thinking he really is kind of silly with his campy hood and sickle and then *bang*, the heads roll. And your systems get all flooey and you're not hungry at all and then you're ravenous and you zone out in front of the TV and then suddenly you're on a tear, raging about one thing or another.

And say that you have a friend who has just begun going through a hospital nightmare of her own and sends you e-mails about it every day. In my case that would be Evelyn, who arrived in Baltimore with Jocelyn and David a few weeks ago and has been pouring out her story anywhere she can ever since. Along with personal notes and phone calls, she sends along long newsletters that she has begun writing to keep her fund-raising troops energized. She writes about the stack of pillows that helped Jocelyn through the long flight to America, the confusion of arrival and the move into the Pierre House, pouring out detail almost like someone trying to numb herself through the sheer volume of facts. She's also full of praise for Jocelyn's bravery and the smile that never fades and how, despite her new neck brace, she manages to eat by shredding her hamburger and pushing the scraps through the tiny opening. The only time she stints is when she's telling Jocelyn's medical details, which she lays out in the flattest tone possible. "The first operation will be on her back and will be about 14 hours in duration starting at about 7:45 A.M. (9:45 P.M. in Australia)."

Evelyn takes particular interest in the other families sharing quarters with them, people who are going through "surgeries that never last less than ten hours" and also missing their homes and also sleeping next to their wounded child week after week. Turning her mind to their troubles distracts her and fills her with tender feelings—there's a woman from California stuck there till Christmas all alone with her sick child and a man from Germany wrenching himself away from his wife and sick child to go back home to make the money to keep the doctors working. "Please keep these people in your prayers," she urges. "It is an extremely stressful and lonely existence."

Once again, getting involved in Evelyn's trouble distracts me from my own. And getting involved in the troubles of Evelyn's housemates takes it even further, until our problems seem to be echoed and amplified by all the sick people in the world. But also, oddly, muffled by them; Evelyn offers so much sympathy and advice and tells so many poignant stories and generally keeps so busy being compassionate that she lets out only hints of the deeper tensions in her own life. Little by little, I notice how rarely she mentions her husband. Or that whenever she does, it's with a snippish tone. "David just left to catch up on his beauty sleep," she writes one day.

She also has some kind of squabble with Dr. Michael Ain, the surgeon scheduled to work with Dr. Carson during Jocelyn's surgery. Apparently she questioned whether he had the stamina to do such a lengthy operation, given his size—he's a dwarf—and he took it personally. But it's not her business to make him feel better, she says. It's her business to make sure Jocelyn will get through all this without any more horrible mistakes. To her, he's just "another arrogant doctor."

By the day of the surgery, Evelyn is so addicted to sharing her story that she takes her laptop computer into the waiting room, typing up bulletins throughout the day:

> *We left home at 5 a.m. and arrived in plenty of time—had a discussion with the doctors about last minute details—then David said goodbye and I accompanied Jocelyn into the operating theatre and stayed until she was asleep. Then I cried.*

In the operating room, the surgeons spend the first hour positioning Jocelyn on the operating table, cushioning her legs and arms with pads to prevent bruising and clotting. Then the neuro-monitoring team puts needles in her neck and toes, and the fluoroscopy team sets up the specialized X-ray equipment that will monitor every second of the operation since even a minor slip of the scalpel could cause paralysis or death. And finally Dr. Carson slips on his rubber gloves and heavy lead apron and begins peeling back the lamina—a hard but flexible material that looks like something between bone and muscle—that protects Jocelyn's spinal cord.

It's now 12 noon. The anesthetist has just come in and said everything is fine—there is minimal blood loss at present but he expects that Jocelyn will need a blood transfusion.

While she types, David reads a book. She can't understand that at all. How can he concentrate?

It's 1.30 p.m. and I have eaten my donuts and am drinking my water (the water negates the calories in the donuts!) and the anesthetist has just informed us that everything is fine but Jocelyn needed some blood so she is having her first unit . . .

The time oozes by like something clotted. She and David barely talk.

It's 3.10 p.m. and Dr. Carson has just been to see us—what a wonderful man. He was amazed at how tight her cord was—there was NO ROOM at all. As he was decompressing, the spinal cord was coming out like toothpaste—something he has never experienced!!!

With eight segments of bone removed, Carson turns the operating room over to Dr. Ain. Standing on a stool to make up for his short legs, Ain begins the process of fusing what's left of Jocelyn's spine by drilling eleven sets of screw holes into the pedicles along her spine, twenty-two holes in all. Since the spine itself doesn't show up on the fluoroscope, Ain can't see it. To avoid a fatal nick, he has to rely on his experience and knowledge of anatomy. Watching the monitor, sweating in his lead apron, he carefully probes each screw hole with a piece of metal to make sure it hasn't gone through the bone into vital tissue.

The last part of the surgery is taking its time. The nurse has just come out again and now the operation is expected to last to 1:00 a.m. in the morning. Neither David or I can sleep—so we continue our walks up and down the corridor.

Many hours later, when the fusion is finally finished, Ain still has to harvest some bone from Jocelyn's hip and decorticate the surface of the spine, which means scraping the bone with a metal tool and throwing in a few bone chips to stimulate the growth of new bone. No one knows exactly why this works, but it does. All this takes more time.

> *Two hours later and we are anticipating an end to the waiting BUT NO, with the nurse's next visit yet another 2 hours expected so now more likely to be about 3:00 a.m. in the morning.*

And then finally Ain and his nurses stitch up the long wound and it's done.

> *Hallelujah, a visit from Intensive Care! The nurse tells us that Jocelyn is coming to them around 3:30 p.m. so an end is now in sight—some 20 marathon hours later. We were the first ones in this waiting room and now we are the last lonely two people sitting in here with my computer, David's book and the overhead music—but we are doing OK!*

After sleeping for just a few hours, Evelyn logs right back onto the Internet and finds a host of messages waiting. At the sight of them all appearing in her message box, one after another popping up like soldiers to the rescue, a flood of warmth washes over her. She hits "reply" and with a single click cc's her legion of Internet friends and fund-raisers. "Hi to all you wonderful supporters! Thank goodness for technology—we have had so many 'visitors' today with support and good wishes, all we are missing is the flowers—NO, that is not a hint!" Plunging right back into the Story of Jocelyn, she tells how her "two slits for eyes" are opening wider with each passing minute and describes the wound that runs from the base of her spine to her neck and exactly how swollen the ventilation tube left her poor tongue. It's another amazing moment of communion and transcendence, another small turning point in the series of turning points that began on that first flight to Atlanta, like stepping up to the ramp at the check-in desk at the Atlanta Marriott and realizing there's another world where difference isn't a tragedy but an opportunity for a new way of thinking about life. Typing on the computer gives her the perfect combination of escape and self-expression—especially since she's finding David's presence increasingly irritating. When he tries to hug her, she pushes him away and goes back to the keyboard. And when she tries to explain that all she can think about is Jocelyn, that she's blocking everything else out, he blows up. "If you tell me one more time you're focused, I'll hit the roof!"

Some of this I learn much later, from David. But lately Evelyn has been letting more and more of it slip into her private notes. "David is going home on the 13th of this month," she writes one day, "and I for one will be glad to see him go. He is not a paternal father and doesn't 'see' what needs to be done."

It's just the strain of the hospital, I assume. When it's all over and Jocelyn is better, they'll find their way back together.

Then the doctors tell them that Jocelyn is going to need another operation. Maybe two.

. . .

And here's the odd coincidence—while all this is going on, Andrea takes my advice and starts visiting her father in the hospital. And I'm visiting my father too, listening to the old stories one more time, maybe for the last time. So we're all in different hospitals together, trading phone calls and e-mails and sharing (as the AA folks say) "our experience, strength and hope." And one day Andrea tells me that my input "has brought a different kind of relationship" between her and her father and once again, it's very moving for me, very satisfying, a taste of what Evelyn is going through with the other mothers down in Baltimore. Even though there's no way to redeem what my father is going through, even though the very thought of trying to find a "silver lining" seems garish, it helps. There's a richness to these hospital friendships that's hard to find in normal life. Andrea says, for example, how very unpleasant it is to sit with her father in silence because his silence reminds her of the past, when the silence would build and build until he lashed out in a verbal assault or stormed out of the house. And silence was also a sign that he was depressed and she hated that so she would work and work to try to get him to talk. And now she's back in the same dynamic, sitting beside his bed with her feet on the stool she takes with her everywhere she goes and chatting away, and when he doesn't talk back there she is right back at childhood with the old feelings, convinced he finds her "boring or stupid." And I write back, telling about my own father troubles past and present and how I've come to see the awkward twisted love behind my father's criticism and distance and even to feel grateful that in his crisis he can depend on me and I can "be there for him," as they say. And there's a feeling of getting somewhere, of breaking down walls.

But there are some squirmy feelings behind those walls, it turns out. Even as these shared experiences bring us together, minor problems keep flaring into big fights. Andrea doesn't like dashing off notes on the Internet and says that writing is too impersonal. She prefers talking on the telephone. But she can't talk at work and I don't want to do it at night, when my wife and kids need attention. This squabble goes on for weeks until I give in and we manage a few calls after dinner. Then everything's pretty much okay until the thoughtless morning when I sit down at the computer and toss off a quick sketch on a visit to a local mall and how disgusting I found all my fellow Americans, bursting out of their spandex pants and "I'm with Stupid" T-shirts. Immediately Andrea fires back an eloquent defense of the common man, quizzing me on exactly why I feel the need to sneer at people for the way they look. There's a hint of alarm in her tone that makes it clear what she's really worried about—that I'm secretly a mean-spirited pig and she's made a horrible mistake trusting me with all her personal secrets. It's the same old problem, just like that first night when she stalked off to the elevator and wouldn't even look me in the eye. And the minute we get over that, I make a remark about how much I hate television and she shoots back a note saying she grew up without a television because her father was a snob (like certain people) and later she learned to *love* TV, thank you very much. And when I respond that maybe perhaps it's possible that she's being just a tad combative, even a touch paranoid, she gets angry with me for making "personal observations." That makes her *very* uncomfortable. It's a boundary issue, and not quite so thrilling to cross as it seemed at first. So finally I just give up and try to keep my notes as neutral as possible—and she confronts me on that too:

> *I'm a little confused. I was thinking that you were going to contact me so we could figure out if we can find a compromise on what we do when I've been offended by something you say about me. Your last e-mail made no reference to that, so I don't know if it's that because you don't*

want to talk about this now, or if you no longer want to talk about it period. For me, continuing to be in touch requires that we find that compromise.

This strikes me as confrontational and uncompromising and basically just makes me want to throw up my hands and scream. She's putting the dwarf mojo on me, dragging me around by a leash of guilt. But then I tell myself that what the hell, she's got issues, and I'm not always the most tactful guy in the world. Patience.

. . .

For Evelyn, these last few weeks have been more and more emotional. One day she gets a check from some people down in Atlanta she barely knows and another day she gets a visit (complete with balloons and a teddy bear) from a Baltimore Rotarian who just happened across her e-mails on the Internet. Another new friend comes by with stacks and stacks of get-well letters and late at night, when Jocelyn's pain gets bad, Evelyn pulls out the letters and reads them aloud. "Hello! My name is Michelle and I am also a dwarf like you. I was very pleased when my principal asked me to do some fund-raising for your operation. It felt good to be helping someone who is like me."

In this circle of sympathy, David is the odd man out. The night before he has to fly back to his job, he lingers in the hospital room. All three of them are tense because Jocelyn's second surgery is the next day and he has to run for the airport the minute it's over. Just then someone comes by with cookies and Sarah's mum comes by with a guardian angel pin and tears are shed and finally they're alone again and Evelyn just tells David he should go rest up for the trip—she'll go sit with Jocelyn. The truth is, that's what she'd rather be doing. How could she possibly tell him all the things she's feeling? He'd just want to charge in and try to fix her like she was some kind of machine he could tune up by twisting a few bolts and plugs. He'd tell her to get some rest or go for a walk. He was never the sort to sit down and have a "deep-and-meaningful," as Evelyn puts it. These last six weeks she's needed support so much and she didn't get it from him; she got it from strangers, from Rotarians on the Internet and dwarfs from Atlanta and Sarah's mum and that writer guy in New York. But you can hold only so much inside and suddenly—standing in that musty bedroom in the Pierre House—she blurts out the question that's been growing in her mind all this last month: "In six weeks, when you see me again at the airport, will you act as if you are happy to see me?"

David frowns, puzzled. As far as he's concerned, she's the distant one. Every time he tries to reach out to her, she pushes him away. And when he pulls back to give her the space she seems to need, she gets mad at him. But this is no time for that conversation, he decides. She needs to be strong for Jocelyn. "Of course I will," he says.

Evelyn looks at him in disgust. He didn't understand at all. He didn't even *hear* her.

After David leaves, Evelyn turns on her laptop and writes to me. I'm the one who gives sympathy with no complications, no demands, the omniscient eye at the other end of her laptop. That's what she needs now and I'm sunk in my own hospital mania and all too happy to provide it. "Tomorrow is D-day again and it doesn't get easier the second time around," she writes. "Jocelyn is a little nervous and on edge but OK."

She doesn't mention the scene with David.

. . .

Andrea's back at the hospital too, looking at her father on his sickbed and thinking that the real problem is that her father doesn't really *like* her. Maybe he loves her, but he doesn't like her and doesn't really enjoy her and for that matter didn't like any of his children and it's not because she's a dwarf. It's something in *him*. And even though this insight is very painful, it is also liberating. She stops being shocked by his little cruelties and heartless precision and somehow that makes it easier for her to sit with him there in the hospital. She's the only one who visits him. Her sisters don't, even his new wife doesn't. And when he tries to be nice to her and say soothing things, she feels nothing. Nothing but ice. He doesn't *know* her. He never even came close.

The next day, she calls me. I'm just back from the hospital myself and not in the mood for a long conversation, so I dodge the call. Two days pass and then she sends me an e-mail:

You've become a disembodied voice. I've lost the feeling of you as a person. I don't hear you, I don't see you, you have no dimension. E-mail now just feels like information, it doesn't communicate personality, god there's no fucking dialogue—it just doesn't suffice.

Maybe it's a guy thing, maybe a moral failure, but right now I feel so unbalanced by disease and hospitals that I need to get a little separate so I can find my strength. Like David. And the fact is, Andrea's need makes me uncomfortable. The connection she seems to have made between her remote father and my "disembodied voice" no longer seems quite so full of healing potential. I write back explaining that I have two children and a job and a father in the hospital and I'm happy to respond to e-mail but these hour-long phone calls are getting to be a burden.

. . .

Jocelyn's second surgery goes well, and Dr. Carson tells them they've finally seen the last of the headaches and drowsiness and memory loss. But Evelyn slips into a depression. It's raining and cold and bleak and Jocelyn is in pain and tired and frustrated too, aching from the surgical staples the doctors left inside her. "She has 30 staples in her head and 70 staples in her back—one continuous zipper!" As the days go by, they fall into routines, medication at eleven at night and again eight hours later. In between they sleep, unless someone makes noise (and someone always makes noise). When Jocelyn wakes up needing to pee, it's five minutes putting on her back brace and twenty minutes getting to the bathroom and back and another five minutes taking off the brace. Then her suture lines start to itch. She's plagued by muscle spasms, too, so Evelyn massages her feet and hands three times a day. She's constantly pale from blood loss and pain. She tires easily.

And the days drag on.

"Please don't forget us now that the crisis is over," Evelyn pleads.

. . .

Andrea writes:

I'm unclear as to what options you're offering, if any. You wrote that you have a preference for e-mail. Why? Does that mean phone calls are sometimes possible? You said I could call when I am 'boxed up and need to talk' and that you'd let me know if it doesn't work for you, but I don't want to call only when I'm upset. I want to call when things are fine, or things just are. I just want you to have dimension.

. . .

One night, sick of using a bedpan, Jocelyn begs her mum to help her to the bathroom. So Evelyn muscles her into the wheelchair and gathers up the IV bag and pain dripper and herds the whole mess across the room. When she finally gets Jocelyn onto the toilet, Jocelyn looks up at her and says, "You're the best, Mum."

Then rehab begins. Every morning at 8:30, Jocelyn reports to Sister Celeste, the no-nonsense nun who has run Dr. Kopits's physical therapy unit for the last twenty years and who knows exactly what a dwarf like Jocelyn can do and can't do—none of that nonsense of trying to straighten arms that can't straighten or exercises tailored to average-size people and simply "shortened." Under a sign that says NO WHINING, Sister Celeste teaches Jocelyn how to use a long stick with hooks to pull on and off her clothes, and introduces her to the miracle of elasticized shoelaces, wiggly wormy things that Jocelyn can grab with her "reacher" and tug tight. By the end of the session Evelyn is weeping. All those years of struggle, of failure, of compromise—and such simple solutions!

Then Jocelyn announces a new goal. She wants to learn to drive.

. . .

Andrea says it's not always my fault. Sometimes it's something in her that probably has nothing to do with me. "I do agree that I have a problem here. I am pretty insecure in this friendship, and I don't completely understand why. For some reason this friendship taps into some crappy feelings."

What can I say to that? How can I spurn such raw honesty? What kind of guy would I be then? So I give in, telling her that dinner is at six and then it's family time and getting the kids to bed, but during the day I usually need a break every few hours. What about we talk then?

No, she says. That won't work. As she's said before, she has no privacy at the office. "It's inconvenient for you at night, it's inconvenient for me during the day. Maybe this friendship is just not going to work."

. . .

When Jocelyn starts to feel better, she puts in a long-distance call to her best friend. Melinda isn't a dwarf, but she does use a wheelchair, plus she's a chatterbox who is happy to fill the long silences—an important quality if you want to be friends with Jocelyn. But this time Jocelyn chirps on and on about the operation and how her hair was shaved off and sharing a room with her mother and doing rehab with Sister Celeste, and Melinda just can't believe it. Jocelyn is coming out of her shell!

Then they get the bad news. Dr. Kopits says the improvement in Jocelyn's spine is causing her legs to twist quite badly. She's going to need another brace—at the cost of $1,300—and more surgery too. In fact, quite a bit more surgery: both hips, both knees, and probably both ankles.

. . .

Andrea writes to say that maybe phone calls at work are okay, if I don't mind her hopping off when business comes up. My response is a little cold. I mention how uncompromising she is and end with a kind of verbal shrug, telling her to go ahead and send her work number. A few days later, another e-mail comes. "So, are you angry with me, giving up on me? It's hard to tell for sure from e-mail."

We talk once or twice and then for no particular reason beyond general overwork, I let some more time slip by without contacting her. Another note appears in my inbox. "I have no idea what's happening between us," she says. "Maybe this friend thing just isn't going to work. I need some live contact."

Another phone call, another week or two without contact, and she writes again:

Our history is one of calls not happening unless we've made definite plans. I'll leave you a message when I get home and know I'll be here for at least a couple of hours.

And again:

I thought I might hear from you this weekend. I don't know if I'm supposed to get that you're dumping our friendship or if you're still busy. If it's the former, please state that explicitly. If it's the latter, can you please tell me a date when you're going to call?

. . .

At dinner, Jocelyn and Evelyn meet a family with a sick three-year-old. The doctors have been doing tests all day and think it's something called ataxia, a very rare disease that will slowly waste her muscles to threads. Then she'll die. The family is stunned, shocked, numb but not numb enough, talking and not talking and crying and not crying. Listening to them, Evelyn thinks back to the day she learned Jocelyn was a dwarf and left the hospital crying so hard she turned on the windshield wipers. Then the grandmother throws out one of those old homilies that are supposed to put things in perspective: "But there is always someone worse off than us," she says.

Evelyn has heard this so many times and said it too, but suddenly she sees it differently—what it really means is, Bottle up your pain, hide your feelings, don't feel so special. WRONG WRONG WRONG WRONG WRONG. This grief is *valid.* She apologizes in advance if she's about to offend anyone and tells them to *acknowledge* their grief. They do *not* have to carry other people's emotional baggage at a time when they are barely able to deal with their own! And the mother who has been sitting there quietly sobbing looks at her and says "thank you" with her eyes.

It's another priceless moment, transcendence and communion all at the same time. And once again, Evelyn feels certain these are her people and this is the purpose of all the pain.

Our [illegible] answering machine. [illegible] alone. I'll never forget the message when [illegible] a [illegible].

Seven

MY SISTER USED TO BE SLIM AND VERY BEAUTIFUL. Then she moved to the Midwest, and the next time I saw her she'd blown up to over three hundred pounds. So one day I find myself reading a brilliant little philosophical antidiet book called *Eat Fat*, which argues that our fear of fat and loathing of fat and general obsession with fat is exactly what makes us fat, and the more we can see the beauty in fat, the less fat we will become. It's very clever and postmodern and a lot of fun to read, with plenty of interesting parallels to dwarfism. Then I come across this passage:

> *Imagine the complexity of feelings when you're really fat, and you find yourself to be the object of the fascinated attention of someone who is thin. And then think how it hurts when you realize that the fascination—the excited curiosity and appreciation of your fat—arises out of some deep revulsion. Naturally, such relationships are brief. The fat woman*

immediately starts to feel like her fat is exotic, like some rare bird, desirable only because unfamiliar, a beautiful ugliness, an alluring freak. It's the same discomfort that arises when whites take to praising the tawny beauty of blacks. The fascination is the form that a certain repulsion assumes in order to conceal its fear.

I read it again and cringe: *The fascination is the form that a certain repulsion assumes in order to conceal its fear.*

I decide to try to be nicer to Andrea.

Meanwhile, unknown to me at the time, a strange double version of my confusion plays out just a few miles away. It seems that even after meeting Meredith at the convention, contrary to what he told me at the time, Michael continued to correspond with Carrie, the woman he had that long Internet romance with, the one he took on-line shopping and thought about marrying before Martha broke them up that fateful day in the chatroom. As he explains it to me months later, he still had some feelings for her and wasn't sure if the thing with Meredith was really going to work out. But somehow Meredith got wind of it and poked around Michael's AOL account and found dozens and dozens of letters from Carrie. And right around the time I start reading *Eat Fat*, she prints them out and goes storming over to Radio City—where Michael is appearing in the Christmas show—and hands him the letters and demands an explanation. They have a big dramatic tearful scene right there on the sidewalk with dozens of New Yorkers standing around in a circle watching, a detail Michael seems to find both horrifying and oddly satisfying.

Eventually, he wins her back. But now Meredith is determined to find out who Carrie is in real life. She already has a pretty good suspect, and with that in mind she studies Carrie's letters and researches her e-mail address and puts together the evidence until she's sure. And what a bombshell it is: *Carrie is Kay.* The big-eyed girl from Texas who used to be fat, who said Michael was her "best little person friend." She was writing Michael notes under this secret phony name and then getting on the phone and pretending to be his friend and giving him advice on how to sweet-talk Carrie.

Much later, I hear Kay's side of the story. Back in 1995, when they were first getting to know each other, she fell for Michael so hard that she pursued him all the way to California. But after a long listless weekend, she flew back home in a state of misery. A week later, on the phone, he told her there just wasn't any chemistry. She was too quiet, for one thing. "I said, 'Yeah, I'm quiet at first, but just give me a chance.' " After that she visited him a few more times, but it never went beyond friendship. And sometimes Michael was *so* critical. Like he'd tell her to wear red lipstick instead of the fuchsia lipstick she favored, and hated it when she permed her hair. "One time at the airport, he said, 'When you talk, you close your eyes. Don't do it.' So I tried not to and people would say, 'What are you doing?' And I'd say, 'I'm trying to stop a bad habit of mine.' It was unbelievable!" He joked that he was Henry Higgins trying to make her into Eliza Doolittle, but she started to resent it and eventually decided it had more to do with him than with her. "I think he was so critical of himself, he'd push it off on me."

But she kept hoping. Then Michael met a woman who was very pretty and very young and Kay panicked, positive she'd never be able to compete. That was around 1996, when the Internet dwarf chatrooms were just getting started, and somehow she got the idea of inventing a girl who was exactly what she thought Michael wanted. That way she could at least correspond with him. And maybe, to be honest, there was some resentment behind it. He always insisted on keeping their friendship a secret, even made her promise not to tell anyone about her visits to him. He said they just wouldn't understand, but she feared the truth was that she was too ugly and he was ashamed of her. Without knowing it, Michael held the power of judgment over her, judgment amplified by love and the fear of not deserving that love. Which left her mesmerized by the horrible possibilities, fascinated to the point of obsession. So she told him about this girl from Florida, Carrie, who was talkative and cool and everything he seemed to want. Michael thought Kay was too close to her family, so she made Carrie an only child whose parents neglected her. Michael liked money, so Carrie was a bookkeeper. For romance, she gave her a night job as a piano player in a bar. And she made Carrie confident and talkative and gave her a hot little body, with blond hair and green eyes and breasts just a little smaller than her own. And made her a bit of a flirt. "Every time I would be that way, it would be, 'You're like a sister to me, Kay.' He had pegged me so I couldn't get out of that mode, but with her I could. I could start over."

And so Michael started writing to her—well, not to her, to Carrie, but it was something. And one day by chance she was on-line as Carrie and Michael sent her an instant message. They had a fairly ordinary conversation, but the minute he went off-line, he called her—called Kay—and told her all about this great new girl. "He was so excited," Kay remembers, her voice still wistful years later. "I thought this is great, he's calling me. That's how it started. It was like a drug, I was so high."

She never expected it to go so far, but soon they were writing e-mails every day and having Internet chat sessions two or three times a week. And after each chat session, Michael would call Kay and go over every detail. When he was on trips, he would ask Kay to write to Carrie for him. And the weird thing is, she'd do it—write to herself, to this idealized version of herself, cc'ing the notes to Michael. "I had a terrible time keeping my e-mails apart," she says. "One time I wrote and signed it Carrie instead of Kay and I panicked." He even gave her his password so she could write to Carrie as Michael. "It was like we were in it together and we were going to win this girl over," she says. "And that's when our relationship began to change. I was his friend. Almost as close as Kris, his best friend."

She knew it was wrong and she knew she had to stop, but she just couldn't. Without Carrie, she was afraid she'd lose him. Sometimes she'd make Carrie act really mean to try to chase him away, but it only seemed to make him more interested. Finally the fear and the fascination got so mixed up and twisted around each other, she wasn't sure what she wanted.

Then Atlanta rolled around and Meredith turned up and she thought it was over now for sure—and then when she got back home she found a message from Michael asking her why she didn't show up. That's something she never understood—if he was so happy with Meredith, why the heck was he writing to Carrie?

But she couldn't stop either.

After a few months, Meredith started to figure it out. One day, she was using Michael's computer when Carrie's name popped up on his buddy list. Another time, she and Carrie ran into each other in a chatroom. And finally, she found the cache of letters. Then came the horrible weekend when Michael and Meredith told AOL some story and managed to get Carrie's home address—which was the same as Kay's home address. He called her up and spoke in a quiet voice. "He said, 'Meredith's out by the pool and I found out something. I found out Carrie is you.' "

She tried to explain, but then Meredith got on the phone and let her have it. When she was finished, she gave the phone back to Michael, and he told her he would never speak to her again. "I was crushed," she says in a tone so morose, it might have happened yesterday. "It was like a death."

When she tells me this story, long after it all ended, she still sounds miserable and a little puzzled, like someone who can't quite figure out the sequence of events that led to the horrible accident that cut off her leg.

The funny thing is, she still misses Carrie. "She made me realize that it was okay to be more assertive and say what I think. I started going out more and taking chances. It sounds like I'm nuts, but I wish I had a friend like her—someone who was really with it and knew what she wanted out of life but was supportive of me."

Later, Meredith tells me that it didn't end there, that Kay continued to call them for another six months—and hang up when she heard their voices.

. . .

One day Andrea tells me about talking to a woman at a party and suddenly the woman started making snide remarks about gays, and man did she let her have it:

> *I said maybe there is more than one family makeup, to which she said that's not what it says in the Bible, to which I said well there are other belief systems, to which she said there's only one right one. This view at times astounds me like I'm hearing it for the first time. It's arrogant, it's naïve, it's self-serving, and it's injurious.*

I love her anger. She can be wickedly funny too, telling me a long droll story about negotiating permission with a priest to become godmother to a friend's child. At first, the priest insisted that she agree to raise the kid in the parents' faith, but she said she couldn't do it honestly because she didn't share their faith. He tried a dozen or so different arguments and she refused to back down, finally agreeing to take the kid to the damn church only after she made it *very* clear to the kid that she didn't believe in all that hogwash herself. But my favorite story dates back to her years in junior high, when she was so stubborn and foul-mouthed that she said "fuck you" to her teacher, eventually saying it so many times that he presented her with a posterboard "fuck you" sign and begged her to just hold it up instead of saying it out loud. What a fighter she is! What a bold and fiery person!

But why does she have to fight me?

. . .

There's nothing this murky in my relationship with Evelyn and Jocelyn, which begins to feel like an antidote to all my moral and emotional confusions. Down in Baltimore, working away under that NO WHINING sign, Jocelyn makes such good progress that Dr. Kopits finally clears them to return home and they make reservations for the end of November, three weeks away. Kopits also tells them that the next round of surgery will actually require *seven* operations, a total of sixty hours of surgery, and the strain of each procedure is so severe that they will have to be stretched out over seven weeks. After that Jocelyn will have to spend another twelve weeks in a plaster cast from her waist to her toes—followed by another twelve weeks in rehab. The whole thing will last eight months. The only alternative is an inevitable return to the wheelchair, as well as the possibility of permanent spinal damage.

Temporary relief comes in the form of a visit by some members of the Atlanta LPA, who cook a big spaghetti dinner and take them out on the town and treat them to a round of crab cakes down at Baltimore Harbor. But the surprise is Jocelyn's continuing transformation. After years of silence, sparked by the presence of the Atlanta dwarfs, she talks and talks and talks some more. And when Mary Alice Johnston responds with tales from her life as a little person, like how hard it was getting into a clawfoot tub that almost reached her chin, Jocelyn listens with wide eyes.

Watching them, Evelyn feels . . . well, not exactly jealous. In a way, she's relieved. But she knows she'll never be able to give Jocelyn what Mary Alice can give, that they are going to an invisible place where she cannot go. At one point Mary Alice turns to her and says, You'll always be her mother. I'll just be her dwarf mother.

A few days later, to my great surprise and pleasure, this letter arrives in my inbox:

I am finally going to write, instead of Mum. Last weekend we had the best weekend with Mary-Alice and Pete Johnston, as well as John and Elizabeth and Jack Howe. They come from Atlanta and are part of our family. We went to the science museum and did some shopping and bought eight pairs of curly shoelaces. Mary-Alice and I also got a caricature, which had both our side profiles and us running with a bulldog chasing us. The caption says, 'Now would be a good time to speed it up.' Mum is doing a fair bit of video taping which I keep on falling victim to, as I cannot use the video with my hand. So if you see the videos ever, remember if I look really bad or my hair is messed up, Mum usually looks just as bad if not worse!

. . .

After yet another argument over some minor remark, Andrea says that maybe we just shouldn't get so personal—in fact, from now on please avoid all questions and inquiries about either my family or my dwarfism. I obey and everything is cordial and superficial for a while. Then she reads one of my magazine articles and sends me a long note about it, with an emphasis on how much it reminded her of my relentlessly inquisitive personality. Responding, I can't resist teasing her a little. "Now who's getting personal?"

She fires this back:

I have not asked you to be less personal. One of the things I like about you is that you do share what you think about things—it just sometimes becomes awkward when the thing is me.

When the thing is me.

. . .

During one of these squabbles, Andrea suggests that my character might be improved by a book called *No Pity*, which tells the history of the disability movement. Reading it, the first thing that strikes me is the familiarity of the mood. "We are subverters of the American ideal, just as the poor are betrayers of the American Dream," writes one paraplegic scholar. "We contravene all the values of youth, virility, activity and physical beauty that Americans cherish." It sounds so much like Andrea, so much like the dwarfs who buttonholed me those late last nights in the dregs of the LPA convention. And the thing is, they're right. Because let's face it, what rootin'-tootin', rough-riding, grab-all-the-gusto-you-can American wants the sight of some gimp or retard or munchkin to put a damper on his good time? Of course we want to hide 'em away! Of course we want to keep those wheelchairs out of sight and shunt the deaf and the dumb and the blind and the epileptics off into nursing homes or institutions, and when they need a bath, hell, strip 'em down and line 'em up and hose 'em down—as (the book tells me) we commonly did in large institutions right up till the late 1960s. And back between the two World Wars when eugenics was all the rage, so many of us were convinced we had to purge our breeding stock of bad influences that the Carnegie Foundation funded eugenics research and Alexander Graham Bell fought the use of sign language in fear that signing would bring deaf people together and increase their chances of breeding up vast deaf tribes. We sterilized the mentally retarded so routinely that the early Nazi advocates for racial cleansing envied *us*, and federal funding continued to flow to such sterilization programs right up until 1974, when it was finally stopped by a court order. As late as 1980, thirty-three states had laws forbidding the mentally retarded to marry. The thought of "defectives" engaging in a normal human activity like sex seemed repulsive, and of course we didn't want them mucking up the gene pool. And we definitely didn't want to look at them unless perhaps we could figure out some way to look at them in a way that made us feel better about ourselves, which is why accepted images of the disabled quickly took the form of pathetic little "Tiny Tim" poster children and "supercrips" who heroically overcome their disabilities. That way we could enjoy our feelings of pity ("poor little thing") and admiration ("she's so *brave*"). A 1991 Harris poll quantified these feelings: almost 100 percent of us claim to admire people with severe disabilities; 74 percent admit to feeling pity; 47 percent say they feel a twinge of fear that the same thing might happen to them; and as many as 16 percent admit to feeling anger at the "inconvenience" they cause.

And then all those gimps and feebs and retards and munchkins finally got fed up. Inspired by civil rights and the antiwar protests and the feminist rejection of "anatomy is destiny," they began mounting protests and mobilizing the media and fighting startlingly bitter battles against the paternalistic restrictions of bureaucrats who sometimes seemed as weirdly obsessed with power as Nurse Ratched in *One Flew Over the Cuckoo's Nest.* It's a rousing story, the sixties protest movement most people have never heard about, complete with heroes like Ed Roberts, a paraplegic iron-lung user who fought his way into an on-campus hospital at the University of California at Berkeley and became so successful at finding his way around prejudice and physical and bureaucratic obstructions that other paraplegics began to follow in his wheelchair tracks, setting off a nationwide struggle for "independent living." By 1968, Roberts and his "Rolling Quads" began the fight for curb cuts. By 1970, they started the first of the independent living centers, with thousands more to come all over the country. Around the same time, retarded "self-advocacy" activists started a group called People First, advancing the radical notion that even people with low IQs should have some kind of say in how they live. In New York, a quadriplegic antiwar activist named Judy Heumann sued the New York Board of Education after it denied a teaching certificate on grounds that she'd be a hazard in a fire, a fight that led Heumann to start an activist group called Disabled in Action that led one of the first wheelchair demonstrations in Washington. That began another trend. When the Carter administration balked at enforcing new legislation forbidding discrimination due to disability, three hundred men and women in wheelchairs occupied the Health, Education, and Welfare offices in Washington for two days. Around the same time, nearly two hundred took over the HEW offices in San Francisco in a dramatic standoff that lasted twenty-five days, featuring deaf activists passing messages in sign language and Black Panthers cooking meat loaf. The more militant wheelchair activists began taking sledgehammers to the streets, smashing the curbs that kept them prisoner. In Denver, nineteen members of a group called ADAPT blockaded an intersection to demand wheelchair lifts on public buses. During a protest in Washington, members of ADAPT threw themselves out of their wheelchairs and crawled up the Capitol steps. The protests woke up the politicians, especially the ones who had disabilities themselves or family members with disabilities. In 1975, following a study by Marian Wright Edelman that found some 750,000 handicapped children were being kept out of school because school administrators said it would be too expensive to include them, Congress passed the Education for All Handicapped Children Act, often called "the disability movement's Brown vs. Board of Education." Over the next decade, increased government funding spread independent living centers and other disability programs across the country. The states eased nearly 200,000 physically disabled and retarded men and women out of large institutions and into apartments or group homes. And finally 1990 rolled around, and Congress passed the Americans with Disabilities Act. As he lifted his pen to sign it, even President George Bush (Sr.) flashed a bit of unexpected militance. "Let the shameful wall of exclusion come tumbling down," he said.

And in many ways, lo and behold, it actually did. A man who was evicted from a Broadway theater as a wheelchair fire hazard sued and won. Another sued his former high school for scheduling the school reunion in a place his wheelchair couldn't reach. Parents began to sue school districts to stop them from hiding their disabled kids away in "segregated" special-ed classes. Thousands of lives were improved.

But there were costs, and problems that seemed impossible to solve smashed up against activists who had less and less interest in any kind of compromise. Oregon's attempt to extend health coverage to 120,000 poor people was derailed by disability activists who fought any attempt to limit services to a realistic minimum, setting a precedent that would help scare other states away from "national health" programs. Another battle rose up around the painful question of the abortion of severely defective children, which happens about eight thousand times a year and which one ADAPT member called "a holocaust that is wiping out our tiny brothers and sisters with disabilities." Never mind that some of these kids are born without any brains at all, or any chance of real life—in the eyes of the more militant activists, any judgment or limit put on what kind of life was worth living was Nazism plain and simple. Even New York City's attempt to put public toilets on the street was derailed by disability activists who demanded that every last toilet be wheelchair accessible, no compromise allowed. And gradually the real firebrands began to move from the dream of integration to "disability separatism," a movement sparked by the 1988 protest at the Gallaudet University for the deaf. Sometimes called the "Stonewall" of the disability movement (in reference to the nightclub raid that kicked off the gay rights movement), the protest began when the university hired a hearing president despite student demands for a deaf one. The students erupted in anger, waving DEAF POWER signs and burning effigies of the new president and boycotting classes until they got their way. Soon deaf leaders were complaining that the idea of "mainstreaming" deaf people into the larger society of hearing people was exactly "like trying to solve the race problem by making everybody white." They even denounced attempts to develop scientific cures for hearing loss—like cochlear implants, an electronic gizmo that stimulates the inner ear—as "cultural genocide." How dare anyone suggest there was anything wrong with being deaf? Members of one of the more militant advocacy groups for the blind even fought well-intended aids like beepers at traffic signals and insisted on sitting next to the emergency exit in airplanes, forcing (in one incident) airplane officials to arrest them and drag them off the plane. Others began protesting the Special Olympics, arguing that it was just another form of segregation and "furthers the pity impulse."

All of these changes roiled through the Little World too. Although many dwarfs are not physically disabled, many are, and all of them look so different that they share the social problems of the disabled—studies show that men over six feet tall get starting salaries 12 percent higher than shorter men, and study after study has shown that almost everyone considers tall men more attractive, more accomplished, and more intelligent too. So dwarfs began taking up the cause of "dwarf pride," agitating for accessible telephones and ATM machines and writing letters to newspapers protesting the use of the word "midget." They fought particularly furious battles against a bizarre fad for "dwarf-tossing" that began in Australia and spread to bars across America—

basically, a bunch of drunks putting padding on dwarfs and placing bets on how far they can throw them. On the LPA website, they even caution against the use of the word "disproportionate," arguing that the limbs of dwarfs are "perfectly appropriate" for people with their serving of genes. They use the words "genetic disorder" but never "genetic defect."

No Pity helps me put Andrea into context. I knew that she worked at a disability organization providing services and legal advocacy, and once she mentioned that she was taking part in a protest over access to a building. But I never thought how deep these politics might go or how mixed up it can get in the personal stuff, the grandmother who wanted to hide her away and the cold father. And even now, it will be at least another year before I realize just how deep and dark it goes.

. . .

After months of promises, I finally leave to visit Jocelyn and Evelyn in Baltimore. My wife is surprised. You're going to visit this woman you barely know? I've told her about the e-mails but not in detail. She doesn't know how frequent they are, or how involving, how caught up I am in the drama of Jocelyn and Evelyn. I don't know how to explain to her how inevitable this trip has become. Or how important it is for me to prove, if only to myself, that I'm not a bad guy. At this point, I don't even have the words for dwarfshock or hospital mania or the dwarf mojo Andrea keeps putting on me. Plus Evelyn has gone through such trouble, arranging for friends to come sit with Jocelyn so she won't have to worry or rush ("I am going to be greedy—I want to see you without sharing") and planning my trip with such ferocious efficiency ("There is also a Super 8 motel along Pulaski Highway—cost for 1 bed: $61.78—Phone: 410-327-7801—E-mail: s8motel@aol.com"). I really don't have any choice.

By this time Evelyn and Jocelyn are staying at Children's House, an outpatient residence across the street from Johns Hopkins. When I get there, a nurse buzzes me through the door and sends me up to a room that looks almost exactly like their room at the Atlanta Marriott, twin beds and the same clamshell suitcases exploding with clothes. Jocelyn looks good and seems cheerful enough, though stiff as a scarecrow in the new back brace that clamps her tight from waist to breastbone. But we barely have time to say hello before her mother hustles me off. For the next few hours Evelyn and I wander around Baltimore Harbor while she talks and talks and talks and talks. She's on fire, manic with the need to tell. She reviews the operation and the tragedies of the fellow patients and the little acts of kindness that made it all such a strange blessing, and when our legs get sore we find a table at a seafood restaurant and sit down to talk some more. Then she gets to the thing that seems to be behind this manic rush of words—she's glad David has gone home. She admits it with a kind of bitter relief. Things were bad. He was driving her crazy. Under the pressure of the operations, they'd grown further and further apart. He just wasn't the kind of person who could share things, who understood how to talk in an emotional way. He was removed, aloof, cold. That last night before he left, she put all her feelings into a question: "In six weeks when you next see me, will you hug me? Will you act as if you're happy to see me?" From the way she repeats these words to me, in such a bleak and bitter tone, it's clear that she's already sure what the answer will be.

Again, I figure it must be just a stage. After the hospital time is over, everything will get better.

After dinner, we walk through the harbor until midnight while she talks and I listen. I try to encourage her as best I can. At the door to Children's House, I give her my best good-bye hug. The next morning, I come back to have breakfast with Jocelyn, who seems to be in pretty good spirits considering all she's been through and all that is still to come. I can't help thinking she's a brave little thing.

A week later the e-mails start coming from Australia. Nicholas has "grown heaps" and needs a haircut and seems a little lost, and Alecia has lost weight "due to exams and other pressures," but both of them have learned "independent living skills" and a few things about how hard it is to run a household. As to David, "he has felt the impact of being a single parent and is really pleased we are home." Once again, a few people were insensitive enough to ask them if they enjoyed their holiday. Evelyn doesn't dwell on that, keeping up the cheery tone as best she can. But it is miserably hot. And Jocelyn is sweltering in her sticky brace. And they're both having a little trouble sleeping. And then suddenly Evelyn calls me on the phone to announce that she's contacted the

American consulate for information about immigration. Also, she's learning her way around the Internet and now has her "very own private mailbox," would I feel comfortable addressing all further correspondence directly to her? "I really like the idea of privacy," she explains. So from now on, please send all e-mails and instant messages to "Free."

Eight

NOW YOU'RE GOING TO SEE SOMETHING writers aren't supposed to show, the story behind the story, the barb on the hook of the tale. I wish there was some way to avoid it, but this book is about exposure and I would be a hypocrite not to expose myself too. So I must tell you a few of the sad and ugly things that happen when my article about the dwarf convention finally hits the stands. The trouble begins with a celebration lunch at a Manhattan restaurant. Michael's face is grim and sour as the maître d' leads him to my table. Meredith's right behind him, and she doesn't look too happy either. When I hand him a copy of the magazine, he says it's too late, he picked it up at a newsstand out in Long Island this morning—and the whole goddamn thing is about *him!* These dwarfs are going to go crazy! Those descriptions of dwarfs with big butts and big heads? That line about Mr. Potato Head? They're going to *blame it on him!* And that line about being the best-looking dwarf in the

LPA? That's the stroke of *death*. It doesn't matter that it's clearly my opinion, written in my voice—they're going to *crucify* him for it. And the womanizing thing is going to come up too. They're going to say he was lying in wait and preying on poor Meredith. On top of which he asked me specifically to describe him as a stockbroker and not just an actor and it should be obvious even to the most clueless, smug, everything-handed-to-him-on-a-platter tall guy just how big a difference that makes in the Little World. It's *a deliberate betrayal.*

Meredith seems much calmer, but she's a little ticked off about the passage where I mentioned she had a leg-lengthening operation, when (she says) I knew that she hadn't done anything of the kind.

I'm totally thrown. It's true that I didn't mention Michael's part-time work as a stockbroker, but the idea that people would blame him for my descriptions seems flat-out absurd. As far as Meredith's leg-lengthening goes, I'm sure it's in my notes—it had some kind of Russian name, Alizoff or something, and the doctors did it to her when they were doing some other procedure on her legs. I distinctly remember double-checking it. More to the point, I was certain they both came off as sensitive and perceptive people. I honestly expected them to love it. For the next hour and a half I do my best to placate them, reminding Michael of the scene when he cried at the airport and all the brave and wise things he said. I mention a few ugly comments he let slip that I didn't quote. And by the end of lunch, he does seem a little more relaxed. I walk them out the door and down the sidewalk to their car, and once again the sight of Meredith limping along with her hand in Michael's elbow hits me in a soft spot. On the bustling streets of New York, they seem so small and vulnerable.

A few days later, Meredith sends me a message. "Do you see what's going on????"

She means the dwarfism listserv, the Internet message board where the little people from all over the world post notes to one another. And yes, I've been watching. In the last year, the listserv and the dwarf chatrooms have become more and more vital to the Little World, a way to keep in touch between conventions and especially to air their many gripes—there are more flame wars on the dwarfism listserv than I've seen anywhere else on the Internet. In the weeks

I've been lurking there, one woman posted a rather brave note coming out as a lesbian only to be attacked for writing "a personal ad." Another woman, mother to a brood of mixed-race dwarf children, wrote a furious letter denouncing the LPA as a bunch of racists. Another woman said she was so sick of the strife she was giving up on the LPA altogether. "After many tries, I found it sort of sad that I was more accepted in the outside world than within the LPA." Now it's my turn. The first few messages are positive enough—a woman named Bailey says she made copies of my piece for everyone in her office and looked forward to even more "biting" articles, and someone named Marty said it "hit the nail on the head on several issues without pulling any punches or sugar-coating certain realities unpleasant though they may be." I'm especially cheered by a private note from Martha, who says she's very pleased with how she came off and even got a call from Michael "apologizing profusely for ever having been a jerk toward people like me!"

Then the attacks start. Anthony Soares posts a message saying he found my imagery "callous," then follows it up with a note saying he's shown the piece to an average-size friend and she also found some of the descriptions (like the ones in the first paragraph of this book) "just plain mean." A woman named Deb says she showed it to an average-sized friend too, with the same result. Then a few people try to defend me and that sparks another round of attacks. "Why are some on the list giving the *Esquire* reporter a break on his often silly and cruel descriptions of LP body configurations?!" asks someone named Vita. "There were plenty of other words he could have used—even somewhat clinical descriptions such as 'short-limbed' dwarfisms or 'short-trunk' dwarfisms would have been better. A furor should have gone up from the LP population. God might forgive this reporter, but I sure won't forgive him!" Another poster says not a single tall person she knows would even *think* anything so cruel. Another says my writing is "very white male" and another calls me "heightist" and another even objects to Jocelyn's story because it has "a hint of 'Big Doctor Saves Little Crippled Girl.' " This poster particularly hates my line about the liberating effect of finding humanity in "inappropriate" packages. "Why on earth would it occur to anyone that humanity did not fill the bodies of folks that look different than you? That is the heart and soul of prejudice."

Most of this, I brush off. It's not so easy for Michael, who's in the nightmarish position of seeing his worst fears come true. Just as he predicted, many blame him for the whole article. One person even says the description of big heads and big butts were not my words but his, that he fed them to me and I just wrote them down. "Michael was really dishing out the lines," writes another. "We in Texas are still cleaning our boots." A day or two later, someone signing himself NJLPA sends this pointed barb: "And as always, some took . . . CENTER STAGE."

This hits Michael hard. Extremely hard. He becomes convinced that now and forever, he's going to be the focus of all the hostility in the LPA. All those years of shaking hands in the lobby and organizing athletic leagues seem to evaporate overnight. It becomes a twenty-four-hour obsession, all he talks about. He screams and rants. He can't sleep. As the days pass and he gets up the courage to venture into groups of dwarfs, he senses hostile glances and hears whispers. Even his relationship with Meredith begins to suffer. Later he tells me he can't blame her for pulling back. "I was showing a side of me that would have been scary to anybody." Finally things are so bad he begins seeing a therapist, who tries to convince him that he's responsible only for the things he actually said—pretty much the same thing I've been trying to tell him. It doesn't help at all. I try to make things better by posting a note (and later a letter to the LPA newsletter) taking full responsibility for my descriptions and praising his honesty and courage, but that doesn't help much either. It doesn't matter if the people in the LPA are right or wrong, fair or unfair. They're his world. And his world is shattered.

The hostility begins to get to me too. The nasty postings may have been easy to dismiss at first, but soon a tone of mournful pain sets into the listserv. "I don't like to think of people making fun of my child, or not giving her a chance because she looks different," writes one saddened woman. Another talks about how disturbing it is "to see people we care about described in this manner." Another says the descriptions made her "giggle with self-consciousness." A friendly woman posts a private note to me saying that she's sure my intentions were good, but "using the terminology you have chosen continues to encourage negative stereotypes that are destructive to us, demeaning and just plain hurtful." A woman named Tricia tells the others that she doesn't think it's fair to blame me when I was just trying to tell their story. "He may not think we are beautiful creatures. But beauty is in the eye of the beholder. I think I'm beautiful for the person I am and the things I do. I liked the article and I felt it was a test to see how strong I am and if I could handle such remarks." Another woman shames me with grace. "The people who are upset are probably the ones who know they look 'funny' to the rest of the world but they really don't like to think that. I know I look funny. It's simply reality. But the most important truth in my life is that I'm happy, my husband (6 feet tall and built like a football player) adores me and thinks I'm the most beautiful woman in the world. The 'being funny looking' truth pales in comparison and importance to the 'he loves me' truth." Others try to help me understand, like the woman dwarfed by a rare disease called cystinosis. "I struggle with my career because I am small, I honestly believe that. And my face is full due to the prednisone I take. But you know those beautiful women you have on the front of the magazine—I've always wanted to be that model. I don't have the proper looks. I am differently proportioned as they say. But a lot of men have commented to me that I am beautiful too."

Worst of all are the kind souls who worry about *my* feelings. "Don't be upset by those who say they'll never forgive you," one woman writes in a private note. "Things will soon calm down." And Angela McTate calls me at work, long distance, just to remind me how upset she got the day she saw me taking notes on her tricycle-riding son. "Didn't I warn you about those dwarf mothers?" She laughs and tells me not to worry and asks for an autographed copy of the magazine for her son.

Reading these letters, twenty and thirty and forty of them every day, I can't help worrying—*did* my descriptions cross the line from truth to cruelty? Was my fascination a version of fear, my guilt cousin to a shameful revulsion? And finally it hits me how very difficult it must be for someone like Michael or Andrea or even Jocelyn to risk exposure of any kind—how difficult and how necessary and ultimately how dangerous. Constantly the subject of curiosity, celebrities of difference, they're forced to bare their necks to the eye of the beholder, opening the very wound they're trying to heal. And then I come along and stick my finger in it.

I am the swine to Martha's pearls.

. . .

Weeks go by and I don't hear from Andrea. Then a few more weeks. I want to know how she feels about my piece, but then I figure her silence is her statement. A few weeks later, I open the latest issue of *Esquire* and find this in the letters to the editor section:

I read with mixed reactions Richardson's article on the LPA convention. I attended the convention and was one of the people with whom he spent time. On the one hand, I was amazed at what he understood about the experiences and thoughts of the dwarfs he met; on the other, I was startled that he missed the essential part. What many of us eventually learn, in part from participating in LPA, is that being a dwarf and having a different body is only that—different. Not, as he writes in his conclusion, "wrong" or "inappropriate." His words imply that although a dwarf might be a great person, his or her physical appearance somehow belies that. We are not contradicted by our bodies; one thinks that only if he or she confuses being different with being wrong. On our behalf, I'm sorry he missed that.

After that we don't talk for a long time—weeks, then months. She had to blindside me with a letter to the editor? She couldn't call me up and talk on the phone? When we finally do connect, we inch out of cold superficiality with the tiniest steps. But I'm pretty sure that for both of us it's a lingering defeat, a small human failure. One day she writes to tell me she's planning a visit to the East Coast and it seems obvious she's telling me this for a reason, that she wouldn't mind if I offered to meet her and try to put things right. So I invite her to stay at my house. A few weeks later, she shows up at the train station carrying her little stool. I'm pleased to see that my daughters seem to like her and don't show any sign of even noticing her size—at bedtime my ten-year-old whispers urgently, "Don't forget to show her the mice." And soon we're eating takeout and falling back into the ironic smartass conversational style that first united us.

But in the morning, the old squabble starts up again. "One of the things that disappointed me," she begins, "is that when we first met, that night at the convention, you talked about how our difference made you think about your own difference. I was hoping to see more of that in your piece."

I'd forgotten about that conversation. "But the piece wasn't about me," I say.

"Maybe it should have been," she says.

At lunch she comes back to it. "I guess the thing that bothers me is, How could I be friends with someone who thinks my body looks wrong?"

"But that's what your father thinks," I point out.

"And I don't like him."

She says it with a flat snap, so final and uncompromising it's a little shocking. But I don't want to back down, not if it means lying to her. Somehow this becomes a point of honor. So we sit there on my sunny deck, eating leftover sesame chicken and listening to the birds chirp as I gas on about health and evolution and how it's natural to fear anything that looks like a symptom of disease and how sometimes I even feel a revulsion against the Hassids who wear fur hats in summer because it feels like a rebuke to my world even though I don't like my world (at least half the time) and it might not be completely rational but it's inescapable and not without reason and if I didn't admit it I'd be a goddamn phony. I know I'm going on and on, but she's got me on the defensive. Anyway, who knows what the end result will be, once you choose to tell the pretty lie. "Choices have consequences."

"But I didn't choose my father," she says. "I'm stuck with him."

"But you chose to be my friend. You chose to pursue me all the way here to my sunny deck."

She gives me another stubborn look, like I'm evading the question. What about the body-looks-wrong line, bub?

"As you know," I continue, "it's always been my feeling you're working something out through me. I think it's because I'm like your father—a judgmental guy who is willing to say horrible things to you. And maybe you're here for a reason and what you need to do is finally say 'fuck you' to me and be done with the judgment of your father."

She seems intrigued by that. "So I should say 'fuck you' to you?"

"You're such a pain in the ass," I say.

We grin at each other.

A week or so later she calls me on the phone and in the middle of yet another conversation about our fathers—her dad's starting to lose his memory—she says she needs to settle something.

I sigh heavily, knowing what's coming.

"Do you understand what the experience is for me when you say my body looks wrong?" she continues. "That's such a profoundly harmful thing for me to hear. Imagine if a friend told you that."

Again, I tell her that I was just trying to be honest. "No pity, remember?"

"I mean, I understand that as a first reaction, that people freak out over difference for lots of reasons—fear of their own difference or the unknown. But people get over it. I've asked my friends about this and they ask me, 'Well, did you ask him in a nice way, or were you defensive about it? Did you get angry?' Because they swear they don't feel the same thing. So when you say it now, it makes me wonder—are all my friends lying to me?"

I listen, taking notes on my pad. At first I thought this was a stage we had to get through, but maybe not. Maybe this is less a relationship than some kind of symptom.

She gives me yet another ultimatum. "I need to know if you're committed to this idea, because if you are I think I have to remove myself from this relationship."

With another sigh, I reprise my position. It's not that I don't know the right thing to say. I *know* we're supposed to say we're above superficial things like physical appearance, that beauty is only skin deep. But it's like my sister getting fat. I love her and everything, but sometimes I look at her and go, "Why did she let that happen to herself? That looks *wrong*." I guess if I were Mother Teresa or somebody I could overlook the body and just see the person inside, but I'm not and I don't want to be. I believe that my experience of *this particular body* has influenced every part of who I am, and it's stupid to pretend that I'm just a soul communing with a bunch of other souls.

"Let's talk about *my* body," Andrea says.

"God, you're relentless."

"You say the alternative is just looking at my soul, but that's wrong.

Accepting difference doesn't mean overlooking bodies. It means getting to the point where difference doesn't seem wrong."

"Look, Andrea, damnit—everyone has their notion of the beautiful. Some people like skinny brunettes. Some people like guys with big muscles. And Freud said somewhere that our notion of the beautiful is connected to our sense of the erotic. These things are hardwired when you're a very young person and not so easy to—"

"I'm not asking you to find us *erotic*," she snaps. "I'm asking you not to see us as wrong."

"Maybe 'wrong' is the wrong word."

"It's an awfully big word," she says.

"It's not like I feel revulsion. It's just—"

"Wrong means not *right*. And my body is *right*."

I don't know what to say.

"This is your stuff," she continues. "I'm not saying lots of people don't share it, but it's your stuff, and I can't expose myself to it."

"But wait—let's go back to my point before. Doesn't everyone have preferences, whether it's for blonds or brunettes or chubby people or thin people? There's always a standard of beauty and deviations from that standard. It's simple normative thinking. That doesn't mean that you can't appreciate things outside the norm, can't look at an old woman or a fat person or whatever and see some beauty in them, or that you can't see past the physical to what they're like as human beings. But not to see the surface at all—that's asking a person not to be human. Or to lie. I mean, I admire your tenacity and everything, but I think what you're asking is not *reasonable*." She starts to object again, but I'm on a roll. "You just want me to think *your* way. You're not accepting my difference."

"I am. I'm just saying it's harmful to me."

"But you're pushing things too far. As usual. You always have to hassle me until you force some kind of confrontation and turn this into some kind of drama. A drama of rejection and acceptance. But things aren't that simple. Sometimes you just have to shut up and let the nuances be."

There's a pause as we both shut up.

"That said," I add, "I wish I was different."

She laughs. That's what I like about her. She gets the joke.

"Look," she says, "if you weren't a friend I wouldn't care. But you are a friend. I care what you think. That's why it's harmful for me."

Exasperated, I tease her. "How old are you, anyway?"

"Fuck you! Fuck you!"

"You're so sensitive. I thought by now you'd be tougher."

"I'm sensitive? Fuck you!"

"Say it again if it makes you feel better."

So she says it again, a couple of times. Then, finally, we move on to other topics.

But it doesn't last. "Anyway, I have to make a decision whether I can go on with this," she says.

"I'll be disappointed in you if you chicken out."

"Talk to someone else," she sighs. "I don't want to raise your consciousness."

• • •

Andrea's visit leaves me rattled. I'm sure she's wrong, but the whole exchange makes me feel queasy. So I hit the library and pull down any book I can find with "beauty" in the title, and this is what I learn: Contrary to popular opinion and natural human decency and all that is good and kind and holy, beauty is *not* in the eye of the beholder. With minor cultural variations, beautiful on Wall Street is beautiful in the jungles of New Guinea. A beautiful woman has high cheekbones and large eyes and full lips and *this is true in all cultures and all races regardless of exposure to the media.* Even infants seem to know the rules—at three months old, they will glance past photographs of conventionally ugly people and linger over photographs of conventionally attractive people. Brain research shows that the sight of a good-looking person causes more neurons to fire than the sight of a plain person. That's why the enhancement of beauty through fashion and makeup is as old as mankind and why bushmen in the Kalahari Desert use fat to moisturize their skin *even during famines.*

In a recent book called *Survival of the Prettiest,* Nancy Etcoff gives the psychobiological explanation for all this, arguing that female beauty is little more than the sign of reproductive health. That's why a study of more than ten thousand people all over the world found that cultures with higher rates of illness and disease place even more importance on beauty than healthier cultures, glossy hair and smooth clear skin being markers of good health, ripe red lips and flared childbearing hips are simply the promise of genetic survival. Which explains why appetites flare in the presence of illness, and why absence of a response to beauty is such a strong sign of hopelessness that it has become a standard part of psychological screening tests for depression. Even the long historical preference for "big hair" may be associated with youth and fertility, as hair grows fastest in girls between sixteen and twenty-four and is easily damaged by poor health or diet. In many ways, female beauty is simply the opposite of male and old; women with a waist-to-hip ratio of .8 (the scientific term for what grandpa used to call "va-va-voom") are twice as likely to get pregnant than other women. A small lower face and chin is another beauty marker, perhaps because men have larger jaws than women and because men and old people of both sexes have heavier jaws. And men and old people also have darker skin and thicker bodies, so men prize women with slender figures and light skin. And men have hairy faces, so hair on women's faces is ugly in all cultures. Male beauty also reflects adaptive success, from the obvious assets of size and brawn right down to the size of the penis (which evolves in response to female infidelity, recent research suggests). Sometimes the connections are bizarre, as in the unexpected power of symmetry—men who are symmetrical tend to have more sex and even give women more orgasms than asymmetrical men, and women with symmetrical breasts have more sex than asymmetrical women, perhaps because men who are symmetrical tend to be stronger and healthier and because women are most symmetrical on the day they ovulate. Nature just doesn't seem to like abnormal things. During storms, more birds with unusually short or long wingspans die, and unusual babies in all species have reduced chances of survival. The natural world favors the average so much that one anthropologist even argues that the average and the beautiful are the same thing and that a "face-averaging device" in our brains churns data until we reach that Botticelli ideal. Bizarre as this notion may seem, studies with morphing technology show that when a group of diverse faces are blended into one face, the composite face is much more attractive than the individual faces that went into it. The more faces added, the more attractive the result. There's even a website where you can "breed" faces and watch the lips get fuller and the nose get smaller, until even the age starts to drop. According to this theory, the most beautiful people are just slight exaggerations of the average, offering bigger eyes or longer hair or larger breasts or a more delicate jawline. But take those exaggerations just a little bit further and the primal alarm begins to clang. Etcoff even extends her argument into more neutral forms of beauty, like landscape art—all cultures and races find beauty in large trees, a variety of elevations and views of water, probably because they are markers of safe and fruitful terrain. And if it feels wrong to say all this, feels vulgar and reductive, like I'm some smug reactionary slapping down a rude slab of raw meat, then maybe that's at least partly the legacy of all these past ideas about goodness and justice and God's image, and because beauty is still far more powerful than decent people want to admit. Mothers of more attractive infants spend more time holding and playing with their children than mothers of less attractive babies. Mothers of sick twins show a quick and clear preference for the healthier and more attractive twin. When four hundred teachers were given the same report card with different pictures, all expected the better-looking kids to be more intelligent and likable. Good-looking students tend to get better grades unless tests are standardized. Adults confronted with cruel behavior by small children tend to think the handsome children are having a bad day and the ugly children are headed straight for juvenile prison. More than half of American CEOs and almost all American presidents are well over average height. Starting salaries are significantly higher for tall men. West Point cadets showing handsome and "dominant" features (prominent chin, heavy brow, deep-set eyes, rectangular faces) were far more likely to achieve high rank. Eighty percent of women say they want to marry men who are six feet tall. When going to sperm banks, women overwhelmingly ask for the semen of tall men. The prettiest girls from high school are ten times as likely to get married than the ugliest girls. The better looking a woman is, the better her husband's education and the higher his income. Women are so hardwired to love the big eyes and chubby cheeks and itty-bitty noses of babies that when they see baby pictures their eyes *dilate involuntarily*. People on the street give tall people about two feet of personal space and even more space to beautiful tall people, but cede less than a foot of space to a very short person. In a study where good-looking women and ugly women wait by the side of the road with a flat tire, the good-looking women always get help first. When good-looking women and unattractive women approach a man in a phone booth and ask if they left their dime in the slot, the good-looking women got the dime 87 percent of the time and the ugly women got it only 64 percent. When researchers "forgot" bogus college applications of pretty women and ugly women at an airport (photos were clipped to the paperwork), people were much more likely to mail the pretty women's applications. Inevitably all of this begins to affect a person's self-image, and Etcoff trots out studies showing that unattractive people left waiting in a doctor's office waited nine minutes before looking for attention while attractive people only waited three minutes and twenty seconds and that when men believe they are talking to a beautiful woman on the phone, the woman on the other end of the line becomes more animated and confident—*no matter what she really looks like.* And while I'm in the middle of reading all of this fine theory, my father's lingering heart failure takes a serious turn and I rush home to find him wasted on his bed, withered and shrunk down into himself and so weak he needs two people to help him to the bathroom. And though he's so far beyond youth and beauty that it seems almost sacrilegious to mention them in the same breath, there's something undeniably alien and forbidding in the purple spots and parchment skin, as if his illness had somehow contrived these signs to mark the boundary between living and dying. Andrea insists "we are not contradicted by our bodies," but right now my father's body is contradicting him. Soon it will contradict him completely. And finally I come to understand that the most absurd and sad and human thing is the way we try to dismiss beauty, to sniff away its terrible power with those pious old homilies. Or offer up a "countermyth" of beauty leading us down to darkness, from Eve to Pandora to Narcissus to Helen of Troy. We deny beauty as we deny death, hoping we won't drown.

Nine

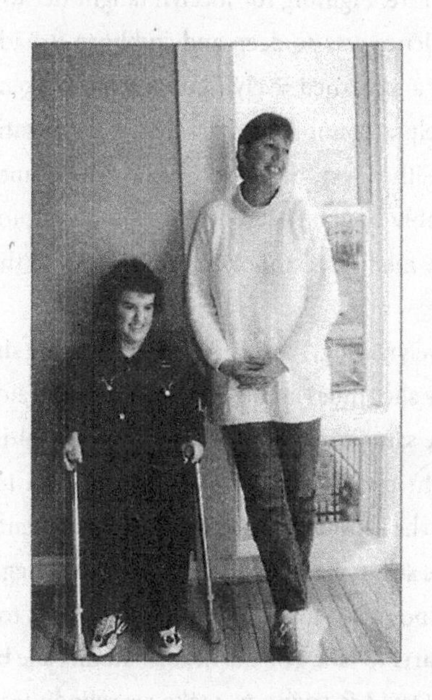

AND SO I FIND MYSELF WALKING THROUGH the frosted double doors of the Sydney Airport into the widespread arms of Evelyn Powell, come because of Andrea and the terrible power of beauty and the angry heartbreak of the dwarfism listserv and because Michael still won't talk to me and because my father died and because of Evelyn's letters, dozens upon dozens of letters that went from strangely noncommittal to deeply depressed to furiously defiant. More and more, I began to see that Evelyn was going through what I'd been going through, that guilt and responsibility and love and the unredeemable brokenness of life were all conspiring to drive her into a desperate state. The first clues were those casual remarks about moving to America. Then she was coming back with or without David. Then suddenly she was moving out of the master bedroom and out of the house and the family was falling into open warfare—accusations, scenes, counselors, lawyers. It seemed to happen so suddenly. For her daughter, Evelyn had wrenched herself out of a sheltered life in suburban Australia and traveled thousands of miles and overcome fears and obstacles. And then went home and realized she wasn't the same person anymore. Fighting for Jocelyn taught her to fight for herself. Or maybe Jocelyn's silence was so deep and stubborn it had somehow acquired categorical weight and turned Evelyn into a kind of agent of her will (which would certainly help account for her ferocious conversational style—she was talking for two). Either way, the consequence was a small domestic disaster that would inevitably bounce back on Jocelyn, a symptom of dwarfism you will not find in any medical textbook but which goes to the heart of this peculiarly social disease.

Wrapped in Evelyn's hug, I'm struck first by how big she is and by the eager and possessive way she smiles at me. She says she came alone so we could talk, and I recognize the same contained pressure I saw in Baltimore, words welling up against the tight-clamped lid of her self-control. I tell her she's looking good and she says she's lost a stone or more, almost twenty pounds, one good thing about all this stress. As we head to drop off my luggage, she keeps to the weather and local points of interest. Then we go down to the famous Sydney Harbor and she starts to talk. For the next six hours, she barely takes a breath. She tells me that David is trying to make everybody feel sorry for him and what really gets her steaming are the hints that she's losing the plot and he even told the kids about her favorite aunt who got depressed toward the end of her life and took Valium and some lithium too, but she was a *wonderful* woman, artistic and eccentric and nothing like her parents who were as strict and righteous and would never let them buy anything on the Sabbath but it was okay for the Sunday paper to be delivered because Dad paid for it during the week! And this idea of moving was never just some fantasy of America because all her life she'd been putting other people first—the family and the children and Jocelyn, which is fine because that's what she wanted to do and had to do, but now she's forty-three and if the surgery goes well, there's a chance for independence and happiness and *something for me*—that's the phrase she's been repeating these last three months in her vast sea of e-mail, "I've finally decided to do something for me." It started not long after she changed her user-name to "Free" and started posting messages about how she really really *really* wanted to become an American like the wonderful open-hearted people she met at the dwarf convention and later in the hospital. "This is not a whim," she wrote one day. "The way I feel about all of this is scary it is so strong." And looking back now, as we rest on a park bench overlooking the water, it's clear that things were already going bad in Baltimore. Before, she and Jocelyn had always done the hospitals together while David was off working. But this time they were together and he was just plain in the way and didn't have the sense to get *out* of the way. He kept telling her to take it easy, which was absolutely *maddening* because she *needed* to stay by Jocelyn's bedside day and night, and if she felt like eating she would eat! She didn't need *him* to tell her to eat. And the pivotal moment was when he was leaving for Australia and she asked him how he would act in six weeks when they met again, if he'd be happy to see her. Pivotal because she knew in her heart what would happen. And it did. When they went walking through the airport doors, there was David and the kids and grandparents and a few friends and everyone rushed up to hug them—everyone except David, who just stood there taking pictures. Later he said he was so shocked that Jocelyn could walk, he didn't know what else to do. And he did cry a little, she did see that. And it's true she didn't tell anyone that Jocelyn could walk because she wanted that to be a surprise. But the point is that his instinct was to hide behind the camera, *even though she'd all but begged him for some sign of affection.* Later he gave her a hug but the damage was done, and then they went home and there was washing to be done and cooking to be done and money to be raised for the next round of operations and it was so hot and Jocelyn was so uncomfortable in that back brace—she would just sweat and sweat until she got this terrible rash. And every day they'd go to one of the local pools for hydrotherapy which took almost three hours there and back, and one day for relief they went to the hairdresser and Jocelyn wanted to dye her hair and it was such an ordeal leaning forward at the basin that she went white and threw up and then threw up again, and Evelyn talked her through it ("Steady on, steady on, we won't let anything happen to you") even though she was scared out of her mind that something *would* happen. And when they finally got home, David interrogated her in a huff about how she could let the situation get so out of control. Ahhhhhhhh!!!! Thank *God* she discovered the chatrooms. That first night when Jocelyn showed her around Yahoo and she just sat there watching the messages scroll by as *complete strangers* talked about *deep intimate feelings* in a way that just made her jaw drop. And when she jumped in with "G'day, I'm a housewife from Australia," the next thing she knew it was two hours later and she had the oddest feeling of waking up out of a dream because the world as she had known it had been exposed as a thin and insubstantial thing shot through with invisible conversations and unknown lives. And the next day she talked to one of the Yahoo crowd again and he showed her how to get into a private "room" and she started telling him—his name was D.C.—some of the stuff she'd been keeping inside about Jocelyn and all her surgeries and moving to America, and suddenly she was confessing her deepest feeling of all, that she was thinking of leaving David—which she'd barely even admitted to herself. It was amazing! She could *communicate* with people. She found herself laughing and joking and realizing that things had gotten so bad she'd forgotten she *could* laugh. Of course David started right in saying she was spending too much time on the computer, so she started getting up at dawn just to type in peace! And staying up past midnight. And she understands it's not just David's fault because lots of families with disabled children have these problems. One person changes and maybe even becomes rather hard because she saw so much pain and shit and developed this tough shell that she doesn't let a lot of people penetrate, and the husband doesn't understand and just wants the fun-loving person he married so many years ago. And David is so unemotional and workaholic and he expects her to work all the time too, and sometimes she just needs a few minutes peace! For God's sake, she grew up with three brothers and was expected to cook, clean, and bottle-wash all day while her brothers did nothing but mow the lawn. And suddenly David had all these *opinions.* She'd organize everything as usual, pushing and fighting, and he's sitting back Lord Almighty dragging his feet and deciding if he approves. Because he was the primary caretaker for Nic and Alecia for all of two months while she and Jocelyn stayed in Baltimore, he's an expert on childcare—"Sorry dear, but that's not the way we did it when you were gone." And then when she tried to involve him, he just said, "Whatever you want to do is fine, dear!" Ahhhhhh!!!! She could *murder* him. Finally she got so frustrated she recruited the kids and they spent two weeks scouring the bedrooms and cupboards and floors and carpets and when that was done, she went to work in the garden, pulling out wheelbarrow after wheelbarrow of weeds and brush and with each weed she pulled, she thought, What to do? What to do? What to do? And then one morning at the breakfast table, David repeated a remark a "friend" made: "Evelyn has changed since last I saw her. She has become much harder. She needs to keep in mind that she also has a soft side." And Evelyn's mouth fell open. How was she supposed to respond to *that?* What did he want her to do, dress in pink? Then it was late January and Alecia had her eighteenth birthday and Evelyn gave her a parachute jump and stood there on the airfield watching her daughter float down through the sky and thinking of all the years that had passed since that blond speck was in her tummy. And somehow that did it. A few days later, she started sleeping in the spare room. And one day shortly afterwards, she and Jocelyn were driving in the car and she told her, "I'm going to leave your father. It has nothing to do with you. It's something happening between adults. So if you ever hear anyone say it was your fault because Mum was spending too much time with you or whatever, don't pay any attention." She spent a long time trying to make it clear that she had come to her decision to move to America completely independently of Jocelyn's decision to move to America. There was absolutely no connection. And when she was finished, Jocelyn said, "I hope you don't get a divorce." And that same night she and David sat the kids down and told them—or rather, *she* told them, while David burned holes through her with his eyes. Alecia seemed confused and Nicholas got red-eyed, and it was awful watching their faces. The next day David took the kids to school and Evelyn didn't even try to fight it. She felt strangely passive, an observer of her own life. She spent the day on the Internet and let David deal with everything and next day the same. They were all so angry at her. Even Jocelyn was angry because David was telling her that Mum was crashing under the strain like her aunt did and using very poor judgment and that the reasonable thing was to have the operation and wait until all this stress and strain was past and then if she still wanted to go back to the United States, he would take her back personally. And in the meantime, if she ever worried about her mum or if anything happened at home that was strange, she should call him at work and he'd rush right home. And Jocelyn mulled that over for a few days and came to Evelyn and said, "Don't ask me where I got this information from, but did you have an aunt that was mentally unstable?" That was a

Monday. On Thursday, Evelyn decided to take Nicholas and Alecia to McDonald's for a deep-and-meaningful. But Alecia balked and asked if Dad knew about this. Evelyn said, "What difference does that make?" and Alecia said, "Because Dad has to know these things." And Evelyn said, "Why?" and Alecia shook her head and said, "No, Mum, we're not going to McDonald's." So they went home and sat around the dining table and Evelyn said—with her heart in her mouth—that she was under the impression they thought she was mentally unstable and that was why she was leaving their father and that hurt *so much.* And maybe because she was always so close to her aunt or because way deep down she'd always been a little afraid of losing the plot herself, but little by little she got louder and louder until she was ranting and Alecia started to cry. Then Alecia pushed away from the table and ran into her room and slammed the door. Evelyn tried to go in after her and Alecia ran out of her room and out of the house and—Oh God! Her daughter was running away from her! And when she collected herself and ran out after her, there was just an empty suburban street and Alecia was gone. Disappeared. And Evelyn got in the car and drove around and still couldn't find her. And when she got back, there was David, waiting outside like a policeman. Alecia had run straight next door and called him. Evelyn got *so* angry then. She said, "Get in the car, I want to talk to you." They drove to town and pulled into a parking lot and Evelyn turned off the engine and said, "*Why* are you telling the kids that I am mentally unstable?" He denied it. They went back and forth for a while and then finally he lost his patience and told her she *was* acting irrationally. "If you begin just with practical matters," he said, "Jocelyn won't be covered by our medical insurance in America. And even if you can get a visa and land a job, you probably won't be able to get coverage for a preexisting condition. So how are you going to live?" And they knocked that around for a while and then he asked, "What was it that set you off? What was the last straw?" And she told him, "It was the day I came home. You didn't welcome me the way I wanted to be welcomed." And he said, "Evelyn, do you realize what you did? You didn't tell me that Jocelyn was on walking sticks! Do you know what that does to a person? When the last time you saw 'em was in a wheelchair, and you're expecting them to come down by wheelchair?" And she said, "I felt no love." And he threw it right back at her. "Yeah. Neither did I." And that was it.

It was over.

By this time, Evelyn and I have left the harbor and wandered up into downtown Sydney. Here among the stone buildings, the sun is getting low and the shadows colder. We find a place to sit down and she tells me how she felt after that final conversation, so alone, more and more alone with each passing day. And suddenly David became this take-charge father. He started spending real quality time with Nicholas and Alecia and they loved it. Now she was the odd one out. And one day he marched into the spare room and informed her that he'd booked them for family counseling. Then he issued a proclamation that he was going to sell the house and move the whole family to America. He said, "I have discussed it with the kids and we are all in agreement." And when they went to the counseling—on the day before their twentieth wedding anniversary—he read out a prepared statement about how much he loved them all and how he thought their marriage had been a real partnership and each of them had strengths and weaknesses and supported each other, and the kids began to cry and the counselor began to cry. Even Evelyn let one tear slide down her cheek for all the years gone by. But then she turned to Alecia and Nicholas and said, "For once in my life I'm doing something for me." And everything got worse and worse and even the Appeal Fund got caught up in the squabble when David convinced them that Jocelyn was "cheating the Australian people" by taking their money and leaving. That was something she could never forgive, because that affected Jocelyn. Because of that, the operation might even have to be delayed. But there was a good part, and that was when Jocelyn came back to her. One sweltering day, the two of them huddled by the air conditioner and Evelyn poured out her loneliness and sadness until both of them were crying and they were a team again. And to keep them going, they set the departure date and picked the perfect day—July 4, Independence Day. The same day the dwarf convention in Atlanta started. But things kept getting worse. By late March, the whole family had deteriorated to an all-time low. Nicholas was rude and abusive, Alecia barely speaking. They spent all their time in their bedrooms or watching TV while Evelyn hid out in the spare room. Finally it got so bad that her brother John and his wife Kitty offered to put them up. And the day they moved out was almost unbearable. Saying good-bye to her children was an agony. She wanted to put everything back in the house and say it was all a mistake. And Alecia and Nicholas were so angry. They weren't even talking to Jocelyn. David threw his wedding ring at her. She just kept saying "Independence Day, Independence Day, Independence Day," and somehow she got through it. Besides clothes, the only thing she took was the family computer and that was a godsend because one day in a chatroom she met a wonderful man named Lee from Texas who was also getting a divorce, and that April she got so lonely she put a plane ticket on her Visa card and flew to Austin to see him in the flesh. Nothing really happened, but they talked all day long, really talked. And every night she went back to her hotel room like a princess in the tower and thought and thought and thought. It was a precious time. But David got suspicious and figured out where she'd gone and went barging over to Kitty and John's house trying to convince them she was unstable—she'd gone halfway around the world to meet some guy she didn't even know, wasn't it obvious she'd lost the plot? He badgered Jocelyn too. Why wouldn't she talk to him? Why was she avoiding him? And of course he told all their friends she was having an affair. And someone sent Lee an e-mail saying, "I know you have two boys, and wouldn't they like to find out what their dad has been up to!" Which she thought was David but turned out to be Alecia, fighting for her dad. Even her best friend Judy told her she was throwing away a twenty-year marriage. But she doesn't care what anybody thinks. The truth is, flying to Texas set her free. It gave her a chance to think and be alone for once in her life and she will *never* be sorry she went—*never!*

All I can do is listen.

That night, back at John and Kitty's house, Evelyn shows me a picture of her latest Internet boyfriend. This one is named Terry. He's black and a bit pudgy, wearing a tuxedo and sticking his hand into his jacket like Napoleon. He looks about fifty. He's a lawyer. Lives in Florida. Evelyn gives the photo a dreamy smile and says she's been talking to him on the phone almost every other day, and even though there have been a few jarring moments—twice he said to her, "I got your photo," when she hadn't sent a photo—she's planning on seeing him very soon. In fact, he's going to come visit her in Atlanta while they're waiting for the operation. For a week.

All I can do is nod.

Then we go to pick up Alecia for a good-bye dinner. She's a beautiful girl, tall and athletic and blond. She's very polite. Around a tanned wrist, she wears a What Would Jesus Do bracelet. Next stop is the supermarket where Nicholas has a part-time job. He clambers in and immediately starts changing out of his checkout clerk uniform into baggy shorts and a Nike T-shirt. Although he wears his hair shaved fashionably high above his ears, with thick bowl-cut bangs, he's also very clean-cut and wholesome-looking.

The jousting with his mother begins immediately. "I want a copy of that report card," Evelyn says. "You can give it to Uncle John, and he'll scan it."

"He's got a scanner," Nic says, like he's really impressed.

"And the grades?"

"I'm top five in science. Top ten in history. Dead last in—"

"What's that?"

"I'm being sarcastic."

"I'm being serious," Evelyn says. Then she gets quiet. "This is the last night we'll be together."

The kids go quiet too.

We eat at a Red Lobster on a crowded road full of fast-food places and gas stations, the three Powells staying fairly subdued and polite and careful of each other's feelings. Every so often Nicholas gets impish and Evelyn gets stern with him, but in the parking lot, when she asks him if he'd like to visit overnight a day or two before she leaves, he doesn't hesitate or joke. "How's Wednesday? How's Thursday?"

They make a date, and he gives her a long hard hug. As he climbs in the backseat, Evelyn looks at me and sighs. "That's the best hug he's given me since this started."

. . .

Here's Jocelyn sitting in my hotel room, serenely unreachable as always. Her mother has gone off to run some errands alone. "So . . . Jocelyn," I say.

"Yeah?"

"You must feel a lot of pressure."

She waits, saying nothing.

"I mean, your mother's done so much for you, and all the upheaval and all."

She shrugs. "Mum didn't do it especially for me. I know if it would have been Nicholas, she would have done it for Nicholas. If it would have been Alecia, she would have done it for Alecia."

When Jocelyn is finished saying what she has to say, she stops. Just as she always does. She never makes idle conversation or social noises. The literature says that this kind of behavior is not unusual in dwarfs, who learn when they are young that a direct and no-nonsense attitude will discourage adults from treating them like babies. But she doesn't hesitate even a second when I ask how David treated her on that fateful day in the airport. "The only time I spoke to him was when he was wheeling me out to the car, after I'd been in the airport for an hour. Everyone else was talking to him besides me."

"What did he say?"

"Something about the weather, how cold it got the night before. Probably a bit about the fly out. There wasn't anything really sentimental."

Then he got her all confused when the trouble started and he accused Mum of deserting Nicholas and Alecia and even of trying to take her, Jocelyn, against her will. He was so upset that for a while, even though she knew taking sides was the worst thing you could do, she started to think he was a bit "badly done by." But then she noticed that he was saying mean things about Mum, but Mum never said mean things about him, and it began to seem "like it was supposed to be Feel Sorry for Dad Day every day." So finally she went back to Mum's side. Then when Mum was in the spare room for those two endless months, he would come into Jocelyn's room at night and tell her his sad story and try to convince her to take his side. So now she thinks it's really all his fault. Even Nicholas's attitude problems are his fault—when they were kids, Mum was always the one who punished them with the wooden spoon, ten whacks on the bare bottom, and Dad barely even talked to them, and that's still the way he is. "Even when he comes home he doesn't even ask, 'Have you done your homework? What did you do today?' He never asks. You always go to Mum for everything. You never go to Dad." And now, after all that, he's accusing her of deserting him too, along with Mum. With a laugh and a shrug, Jocelyn sums up: "He's just an idiot."

"You're tough," I say.

"I know," she smiles. "I don't care. That's how I like to be."

Until today, I've never spent more than an hour alone with Jocelyn. We've exchanged maybe three e-mails and talked a couple of times on the phone. She's always been very cool and adult, and I've always been a bit dismayed by how conservative she is for a teenager. She seems to have no interest in boys, clothes, parties, makeup or pop music, and she can be very harsh in her contempt for human weakness. But over the last year her mother was a bridge I could cross to reach her, and now that we've spent a few hours teasing each other about our terrible accents and getting comfortable, I sense that she's decided to reveal a piece of what she keeps inside—to give me the gift of her secret self. And once she gets started, she really seems to enjoy it. She talks about how angry she is with the Australian doctors for letting her get in such bad shape and the hope she found in America, where people don't stare and even "the transport" is better. "Here there's a big step, and waychurs have to be carried across."

"Carry what across?"

"The whaychur."

"The what?"

"The *whaychur*."

"Speak English, girl."

"I am! You're not!"

She laughs, her face turning red. Then, as it fades back, she tells me there's a split with her brother and sister too. That started when they got back from Baltimore and nobody seemed too interested in all the videotapes Mum had shot of the rehab—all that evidence of how hard she'd worked and how much they'd gone through just lay there on the shelf gathering dust. "That's really meaningful for me," she says. "I would have sat there and watched it with them and explained it, but they didn't want to know about it." Then, with the separation, Nicholas and Alecia took Dad's side. And even though they never blamed her, all the talk of Mum's terrible stress sort of went in that direction, didn't it? And then when they moved to John and Kitty's, there was the blowout with Nicholas over the computer—she started to pack it up and he said it's not yours, it's the family's. As she tells the story, an evil little smile crosses her face. "But I won," she says.

"How?"

"I cried, and Dad came in and yelled at Nicholas." She laughs. "Easily done. And then Mum came and yelled at him—that's even better!"

"At least you know the darkness in your heart," I tease.

She chuckles. "Yup."

Anyway, she says, Nic and Leesha are really just taking Dad's side out of pity. "They feel sorry for him. That's the only reason."

"Maybe they're also angry at your mum," I suggest. "After all, she's leaving them and going to America with you."

"Well, they chose to stay."

"Really?"

"That's their choice."

"You really think so?"

"Well, Mum asked them flat out, 'Do you want to come or do you want to stay?' And they both said, 'Stay.' "

"And you don't mind saying good-bye?"

She shrugs. "I feel a bit sad leaving them. But that's about it."

• • •

That night we drop by the house that David and Evelyn built for Jocelyn when she went into the waychur. It's brick with an Aussie-style tin roof, way out on the fringes of the Sydney sprawl. Inside it's spacious and spare, with flower prints and a grandfather clock.

Alecia brings Evelyn an envelope. "Here's this letter that came for you."

"Who opened it?"

"The bogeyman."

Evelyn shoots me a look, and Alecia glances away. For a while there, Jocelyn was writing to Lee in Texas and calling him "Dad" and one time David caught her and got very upset. Ever since, Alecia has been trying to protect him. She's like the little mother of the house now, Evelyn says, all roles reversed.

Behind us on the kitchen wall, there's a bulletin board slathered with family pictures: Nicholas holding a plastic guitar, Alecia as a pudgy baby, Jocelyn holding a bridal veil and beaming. At that age the only sign of her dwarfism is a bit of extra size to her head, like it was clipped from another photo. "What great pictures," I say.

Silence. It was exactly the wrong remark to make.

The next morning, Jocelyn and I come back so I can talk to the rest of the family and get their take on all that's happened. While Jocelyn waits in the kitchen, Alecia takes me into the TV room and tells me that growing up with Jocelyn was mostly just a normal life in a normal family except every now and then you'd catch a glance and think, gosh, she *is* different. They were close. Primary school was probably the first time she noticed kids staring at Joc, but you get used to that, and nobody ever did anything really mean aside from leaving her out of games and not making friends with her. But they got through that too. Things did get harder as she grew older and started looking really different, and harder still when she started going to the hospital so much and using the wheelchair and having headaches all the time—that would have been in school year eight, the year Jocelyn broke up with her friends, who sort of betrayed her by pretending to be her friend while the teachers were around and then, the minute the teachers left, telling her to go away. Which is maybe why Jocelyn doesn't mind leaving Australia now. But all through all of that, the Powells just hunkered down and got on with it. "I guess the whole family's always been strong about it. I didn't really notice anything different or any strain. We just kept going."

Resentment? Jealousy? Anger? Alecia shows no signs of it. "You've got to look at it in a good way instead of a bad way." She fingers her What Would Jesus Do bracelet. "Like last year when they went to America and I was doing my last year of high school and taking the test for Uni, which is this big stressful thing, most of the time I was focused on studying and my exams and stuff like that. So because I had something else to occupy my time, I didn't really think a lot about that stuff. And knowing that Jocelyn's going to come home, and she's going to be able to walk, and she's going to be able to live a life like *I* have—that helps."

Like other tall people who have passed through the Little World, Alecia seems grateful for the experience. "It's really made me who I am, living with her. Like outside, people may look different, but we're all the same, and some people have just had a really hard time in life, and you just have to feel sorry for them."

When Jocelyn went into the wheelchair, it even helped her find God.

And the divorce? The first hint of trouble was all the photos they'd brought back from Atlanta, so many photos, and it was "have a look at these" and "have a look at these" and all the people looked just like Jocelyn and Jocelyn looked so happy. "I guess she just feels more comfortable with other little people." Alecia doesn't seem disturbed by that, just shrugs it off. "She takes to them straight away, 'cause they're the same. I guess it's the same as females feeling more comfortable with other females, and males with males." As to what happened in Baltimore, she doesn't know what went wrong and doesn't really want to know—all that matters is Mum came home so fixed on moving and kept saying this place isn't home and she had to go live in America, and even if you tried to remind her about all their friends and the people who helped them raise the money for Jocelyn, she would just shrug it off. The closest Alecia gets to a criticism is to say that yes, she does think the surgery had a lot to do with her mum wanting a divorce. "With so much stress, without her knowing it, she's pushed a lot of people away," she says softly. "Because she wanted to deal with it herself."

But why dwell on it? Things happen. Everyone has their life to live. "It hurts, yeah, but you've just got to get on with it."

Alecia's so stolid and strong, she could be standing next to a tractor in one of those old Soviet propaganda posters. But when I ask about her father, she slides forward in her chair and a touch of anxious indignation comes into her voice. "I would like to think that he was a normal father. He wasn't around as much, but he was still my father. When Mum told me, 'Oh, because your father was never around when you were little, he doesn't know how to be a parent,' that really hurt me. Because he was still a parent to me. I still thought about him a lot, and spent as much time with him as I could. I don't know why she thinks that."

I start to ask a question but she rolls on, eager to make her points. "It's true that Mum's car was the one that was always out front, because Dad left early to go to work and never came home until it was dark. And Mum was the one who came to school and did the sports teams. Dad never did that. But he always went to things like parent-teacher night interviews and performances and stuff like that. I know sometimes it looks as though Dad doesn't care, but it's not true. I *know* he was there emotionally. He tried as much as he could. He's not the huggy type, but an emotional person doesn't have to be huggy. An emotional person is someone who has emotions. He showed love. He worried. That's emotion. It's just that Mum is *very* emotional."

She's talking fast now, without a single pause. Clearly these are things she's been wanting to say for a while. "A lot of people think that Dad hasn't been a part of the family, that he hasn't been helping out, and I can tell it hurts him so much. Jocelyn's leaving, and he can't be there when she's in that pain, because Mum's there."

"That mus—"

"I know Mum thinks Dad hasn't had a major part in Jocelyn's life, but I think he has. I mean, I look at Jocelyn, and I see Dad. Jocelyn always got really high marks, and Dad is so smart it's not funny. The way they sort of bottle everything up."

"That comes from your dad?"

"I know Jocelyn's mad at him, but the reasons that I've heard just aren't—they're not reasons."

Fingering her bracelet again, Alecia says that one good result of all this is that her faith has gotten stronger. She started leaning on it when Mum and Dad were in Baltimore, just opening the Bible and looking for guidance. The breakup made her go deeper. "It helps me to understand why Jocelyn causes so much pain," she says.

I'm not sure I heard her right. "Why she *causes* so much pain?"

"Yeah. Well, not *her*. Because she's a dwarf. All the medical problems, and the family, and then, I mean—yeah. Why did that all happen? Why couldn't God just give us a normal sister? And now I know God made Jocelyn the way she is for a purpose. Although my parents did split, I think she's created something special in the whole family. Someone told me that Jocelyn has a special gift, and she's here to give out these gifts to people. She's here for a purpose."

. . .

Sure, Nicholas says, his grades plunged when his parents split up. Big deal. "I was getting to the point where I didn't study for tests, and I was getting nine percent."

He laughs. And last year it was fun, really. When the folks were in

Baltimore, he and Alecia lived on their own and organized transport and paid bills and found out they could do a lot by themselves. It was good for them. Fifteen years old and on his own. It was great, really. So yeah his grades started to slide a bit, but that was only because there was no one there to kick him in the ass. Even when they came back from Baltimore and the trouble began, he wasn't that upset. School had just started. There was some "stuff going on with this chick."

He stops. "Do you know who Puff Daddy is?" When I say yes, he looks skeptical, as if such illicit knowledge were reserved exclusively for teenagers. Then he leans forward and lowers his voice to a whisper, telling a story about Marilyn Manson he thinks will shock me.

When I turn the subject back to his mum, he gives a cool shrug. "I don't know how I feel toward Mum, exactly. There's stuff I think about things. I just think—I don't know, that she could just handle things a bit better."

Like her Internet habit—now *that's* annoying. And the people on the Internet are annoying too, all her anonymous buddies with their helpful advice. "I think that's wrong for someone to give advice to other people to end relationships and that, when they have no idea what they're talking about. And I'm annoyed that Mum did not at least give Dad an advance warning and say, 'Look, I'm feeling this way, I don't know if I want to fix it, but I'm just letting you know.' "

And it was also annoying after she moved out, when she told Nic she loved him and he didn't say anything back, and then she told him again and he still didn't feel like saying anything back—he shouldn't have to say a thing if he doesn't feel like it. So now it's just "hi" and "bye" and nothing too deep and meaningful.

But he doesn't blame Jocelyn. How could he? Even the way she took sides he can't fault because she's stuck in the middle and she has to work with Mum and keep that close relationship going because she has no choice. "She'll be spending every day, every second, every minute with Mum for the next year. So she has to, no matter what."

And really, it's good that Jocelyn wants to start a new life in America. That's her choice. He's just a bit annoyed that they're flying away on Saturday because they don't really have to leave for three months and without being big-headed or anything, isn't a person's sixteenth birthday sort of a big thing? That's why Mum originally planned to leave in August, so she could be there for his birthday. Then she went to Texas, and when she came back she decided she had to leave on the American Independence Day. What the hell for? So he talked to Alecia about it and she said turning sixteen was no big deal but still—every five years they've always had a big birthday party for each kid and last year he didn't get his celebration because of all the fund-raising stuff and this year he was really hoping it was going to be his turn. "I know Jocelyn has so many needs and all that, especially when she started to get physically ill," he says, "I mean, I'm not angry. I can't blame Jocelyn for what she's done, or Mum and Dad for spending more time with her to figure out what's going on. I'm not jealous. I'm not angry. I'd help her if I could as well, but I'm not old enough. It's just that these are family things a mum does. And to me, she's just like totally not being a mother."

. . .

And here's David sitting back in a comfy chair, giving me the man-to-man smile of a guy who has repeated the words "these things happen" so often that he's achieved a numb kind of peace. But the moment I turn on my tape recorder, he comes to attention and starts crying. "Excuse me," he says, pinching his fingers to his eyes. "I'm very upset."

I apologize for the intrusion.

"I don't think Evelyn understands the pain it's causing this family. This is a *hard* time for me, because Jocelyn leaves on Saturday." He's sobbing openly now. "Our relationship is not good at the moment. I think Jocelyn has shut down the communication because of the conflict between me and Evelyn. The only time we talk is when I initiate it. When she leaves, I won't see her now for nine months—nine *months*."

David wipes away the tears and takes a calming breath. He's smaller than his wife and looks exhausted, as if the air has leaked out of him these last six months. He says that Evelyn has always been a hard-driving person. His secret nickname for her is "the Bulldozer." When he wakes up in the morning, he wonders what gear she'll be in today. But he never thought she'd bulldoze through him, never thought she'd rumble right over their marriage. He wipes his eyes again. "I don't know whether there are signals that she picked up on and therefore she decided—but, shit, they're the wrong ones, John. They really are the wrong ones. You don't shut out a person like this. After twenty years of marriage you would sit down and say, 'I've got concerns about this, and I think we should really talk about it.' Even when I said to her, 'This was a *great* marriage we've had,' she went 'pfff.' "

David met Evelyn at a party in Sydney when he was just a young engineer and she was starting her nursing career. It wasn't love at first sight, but there was something he liked about her. She was very reserved and also surprisingly aggressive—after dating him for three months she said she loved him. And when he said he didn't quite love her, she told him that was okay, she'd wait. So they got married, and six months later she became pregnant with Alecia. Then came Jocelyn.

Evelyn was crushed by the way people reacted to Jocelyn, David says. "Some of the family were in a state of shock for a while. Some came round and visited and some didn't, and I know Evelyn was angry with the ones that didn't." That was when she started cutting people off, a habit she continued through the years. She even broke up with her best friend Julie, who had married David's twin brother. Last week David asked Julie about it and she told him what happened—one day Evelyn was talking in her obsessive way about Jocelyn's problems and Julie asked her if maybe the reason she couldn't stop talking about it was because in her heart, she still hadn't accepted Jocelyn. And Evelyn never spoke to her again. Maybe it's the same sort of thing that's going on now, with him. "There's something really driving Evelyn. Does she feel guilty? I don't know. There's something behind it. She was fighting so hard for people to accept Jocelyn, and I think she missed the point—numbers of times I've spoken to her, and I've stipulated that *Evelyn* was the one who didn't accept Jocelyn. When everything happened medically, she was in there, boots and all. But I just got the feeling as a mother that bond wasn't there. Maybe that's why she was so keen to have another child. At the time, I asked her why, and she answered, 'I've got to prove that I can have a normal child.' "

When David says this, I'm appalled. It's so obviously self-serving, the ultimate low blow. Or is it? Could this be why Evelyn has taken things to such extremes? Until Jocelyn was almost eleven, he continues, he was the one who was closest to Jocelyn. Even Evelyn admitted this, he says. "She passed a comment once, 'You and Jocelyn were together for ten years, and then I started to build up a relationship with her.' And that's the way it was. I think I can say Evelyn accepted Jocelyn as a daughter, but I don't think she accepted the fact that she was a dwarf." It was only when Jocelyn started getting that burning sensation in her legs and couldn't walk distances that Evelyn became the perfect mother—and she did, she really did, then. And they got through it. And everything was fine.

Until Atlanta. And the funny thing is, the night before she and Jocelyn left for that fateful first trip to the LPA convention, Evelyn was so scared and cried and cried and kept saying how much she hoped they would grow old together. Then she came back with all those pictures and a colored brochure of all the houses you could buy and raving about how great it was in America. That's when the distancing started. Evelyn said she was just focusing on the charity, but she wasn't showing any emotion at all. By the time they got on the plane to go back to Baltimore together, she would barely talk to him. "I mean—*shit*—I'd just left two kids behind, and I try to talk to her about it, and she says 'Don't talk to me about it! I'm focusing on what we have to do next!' She got up and moved away from me and sat next to Jocelyn. And I thought, What is this?"

The whole time they were in Baltimore, she was icy and remote. "I could not even touch her. Could not really talk to her. Every single thing I said she would misconstrue. My perception was that if I said something, Evelyn *bit*." At one point she asked him why their relationship couldn't be more friendly, and he said, "Evelyn, it can be, anytime you want it to be." And she went, "*Ooouuugh!*" and walked out of the room. A week later he said, "Evelyn, this is nuts. We've been here for a week, and you elected to sleep in the same room as Jocelyn because you didn't want her to be afraid, and I'm totally supportive, but when the room became free which had a double bed and a single bed, and we could be family together, you refused to move into it. You won't let me cuddle you. I'm not saying it's got to be for sex, but just being near you. And to hold you. And you to hold me. We're going to need each other. And her answer? 'Stay away from me; I've got to focus, I've got to focus.' "

That was when she started diving into the computer. Sometimes he'd look over her shoulder and she'd tell him to butt out, that he wasn't allowed to read her mail. "For heaven's sake," he'd say, "he's a journalist from New York! If you want to read it first, fine, but if I can't read it afterward . . ."

The final straw was the day Jocelyn was being operated on the first time and Evelyn came out from the operating room scared out of her mind, saying she was so afraid Jocelyn would die, and when he tried to tell her everything would be okay, she looked at him in shocked surprise, pulled her hand away and just shut down. Who knows why? She won't talk about it. And that thing about the question she asked him the night before he left, when she asked, "When I come back from Baltimore, will you come up and hug me and do da-da-da-da?" What he said was, "Evelyn, of *course* I will!" He didn't think much about it, figured it was just Evelyn needing a sign of affection and security. What did she expect, him to race down the corridor going, "Ooh, darling"? But seeing Jocelyn on the walking sticks hit him like a ton of bricks. He was just shattered. He just kept on taking pictures of her and people meeting her because he wanted to capture the moment and plus he was exhausted from working with the charity and his new job and looking after Nicholas and Alecia. There was also the memory of Baltimore, all those months when Evelyn had been so cold to him.

And that was it. After that day, her mind was made up. Whenever he tried to talk to her reasonably, trying to convince her that she'd only seen America under stress and under those conditions you always meet angels, or that she should be careful because it would take her a while to get her feet back on the ground, she would get angry at him. He even went to see his boss about a sabbatical so he could go to America with her the next time, this time, and his boss suggested he go to North Carolina and work in a branch office for five months. "And they were going to *pay* me." But Evelyn said a sabbatical wouldn't be enough—she wanted to *move* to America.

He starts to cry again.

Then she started that Internet business. Not like in Baltimore the first time. This was way over the top. She'd get up at four in the morning and at nine o'clock she'd still be in her pajamas, typing away. At night she'd get on at eight or nine and go to one or two in the morning. "I said, Hey, take it easy with the Internet. And she would say she was fine, it was just an escape."

Two weeks later, she told him the marriage was over.

Ten

SCIENTISTS HAVE STUDIED THE "ETIOLOGY" OF FAMILIES like the Powells and discovered certain patterns. Almost without variation, parents experience heightened anxiety when they get the first diagnosis of dwarfism. They often find it difficult to accept and go for second and third opinions, hoping to hear something different from a new doctor. They often clash with doctors who "tend to see parental anxiety as interfering with their medical management," as one researcher put it. As they struggle to come to terms with their child, they go through periods of disbelief followed by predictible stages of "anger, denial, avoidance or intellectualization," and also through an interlude of grief over the perfect child that was not born—a grief that tugs them toward isolation. They are often troubled by negative feelings toward their imperfect children and the kind of bulldog anger that rises up to push aside their terrifying feelings of cosmic helplessness. And most of all, they are troubled by guilt. The literature is quite definite about this. "All parents of children with health impairment experience guilt," one review of the pertinent studies emphatically concludes. And they tend to atone for that guilt "by lavishing excessive attention and care upon the child (who is almost certain to become aware that he or she is a burden)." This familiar and universal reflex is especially common among parents with children who have to go through multiple medical procedures and hospitalizations. And since none of this proceeds on a neat upward line from denial to acceptance but in a swirl aggravated by medical and social setbacks, these troubled parents tend to spin back through their guilt and grief and denial and intellectualization and all the rich evasions the heart discovers—including "irritability within the family, displaced feelings of anger deflected from the child onto the spouse and vice versa, critical feelings towards professionals, bitterness, marital conflict, apprehension, and excessive anxiety."

That's what it says in the books I read much later, all the heartbreak couched in language that seems designed less for accuracy than distance. But right now Jocelyn and I are driving away from the house she once called home and the man she once called Dad and she doesn't ask me a single question about what he said. Not one. She doesn't ask what Alecia said either. And I don't say anything because I can't believe she's not curious. I want to see what will happen if I just let the silence grow. I want to test her breaking point. But the silence grows and grows until the one in the car feels like lead, and she doesn't so much as blink. She's so self-contained it's almost comical, like she has more density than the average human, like she's tamped down into herself and that's the reason she's so squat and small—not genetics or some failure in the fibroblast growth receptor but sheer stubborn reticence. Finally I can't take it anymore and ask her why she isn't asking me what happened.

She shrugs. "I figured if you wanted to say something, you would."

. . .

Evelyn rushes to the computer and finds a Universal Resource Locator from Terry, the lawyer who lives in Florida. She clicks on the throbbing little icon and rides it to a cyberbouquet. Cooing at what a dear fellow he is, she starts writing a flowery thank-you note.

"I can't believe you're having this intimate exchange with a man you've never met," I say.

"It's easy, because we communicate," she replies. "We write honestly to each other, and he's everything that David isn't. He's honest, he's direct. He asks point-blank questions, which puts me on the spot, and I have to answer them. And I like that."

Evelyn introduced me to Yahoo and Hotmail and ICQ and sent me my first URL and has always been a little disappointed, I think, that I didn't fall in love with the wide open cyberspaces the way she did. Trying one more time, she stops typing to sum up the appeal. "People on the Internet are honest with each other because there's nothing to be afraid of—you can always just log off. But how do you just say to somebody standing in front of you, 'Exit. Delete.' You can't."

Also, they give her so much support. Without them, she doesn't know how she would have gotten through these last six months. Even now they're helping her. Just look at the messages they sent the last few days. "Enjoy the trip over," one says. "Bring some money," says another. "If you don't have any money that's OK—come anyway." Each message is a small bit of cheer in a world where almost everyone else is completely against her.

Then she turns back to the thank-you note and signs it with three letters: "ILY."

"That's gone a little far," I say.

Her voice goes soft. "It started a while ago."

. . .

Jocelyn's wheelchair needs to be fixed, so we drive together down to the store where she got it made back in 1993. She was excited then, so happy to get off her sore legs. Now she hates the thing and dreams about cutting it up with an acetylene torch.

The lady behind the desk says, "Oh, I saw you in the newspaper. Congratulations."

"We're going to be on TV again," Jocelyn tells me. "We're going back for another operation."

While we wait, I quiz Jocelyn about her family, and she tells me she's mad at Alecia for writing to the man in Texas behind her mother's back. And for telling her dad what Evelyn was saying and doing. And she's mad at Nicholas for ignoring her and her feelings. When I suggest that they're just worried about their father and mention how much he cried with me, her face sets.

"I don't buy it when he cries," she says.

"You don't?"

"No, I don't."

She doesn't elaborate. I'm going to have to push if I want any more details. "Why do you think he cries then?"

"To get the emotional support. Like you'd see him as poorly done by. He needs a lot of help. And a lot of support. And he needs it, not Mum. Which is not right."

"How do you know he doesn't actually feel that way?"

"Well, I think he does a little bit, but I think that stopped a long time ago. And he's just gotten used to the story, and crying every single time he explains it. The first few times, okay, but not to keep going like he does now."

The wheelchair fixed, we drive over to Jocelyn's school. While she waits outside, the principal tells me how they adapted to Jocelyn with curb cuts and sports therapy and an elevator, taking it as part of their Christian mission to respond to her special needs. Then he surprises me by saying straight out how Jocelyn's illness wore them all down—there were days when she would go to the support woman hysterical and crying, and the traumatic break with her circle of friends roiled the school and left them all deeply saddened and dismayed. He has an autistic daughter himself and knows something about the trials and joys of such a child and was always conscious of grabbing the opportunity for grace. But that doesn't make it easy. After an hour, he turns me over to the deputy principal and then to Jocelyn's class adviser, who tell me that Jocelyn's trouble with her friends began almost on the day she started using a push wheelchair between classes and seemed to progress in step as she gave up walking completely and then moved to her electric wheelchair. The other girls just weren't sure how to act anymore, and it was traumatic for them too, seeing her get weaker and weaker. There were nights when Jocelyn went home crying because the other girls were ignoring her. Sometimes she would get testy and that would make things worse, so she became more and more remote. Then came the day when Evelyn stormed the school and said something *had* to be done, so they set Jocelyn up with a counselor and a pastor and tried to get her to treat the special-ed office as her "home away from home." But you can't force teenagers to feel what they don't feel and not to feel what they do feel, and it all came to a head in the sad and pivotal meeting when they summoned Jocelyn and her friends to the office and tried to get them all to talk it out. Instead they talked it deeper. "She sort of shut down very quickly then," he says.

Except with David Jackson, the head of the special-ed program. She visited his house, went to his sleep-away camp, fled to his office when she was in too much pain to sit in class. His own daughter was one of Jocelyn's ex-friends. With a beard hanging down his chest and a watchman's cap on his balding head, he's an appealing mixture of rumpled uncle and Old Testament prophet, and when I ask him how Jocelyn's disability has affected her family, he leans forward in his chair and checks my tape recorder to make certain it's working properly. "I'll speak plainly on the subject," he says. "I don't mind if Evelyn listens to the tape."

Then he leans back and takes a deep breath. "It is the common feature of families with a child with a disability that the mother of the family is the primary caregiver from birth. Mother is the one who has the child on the breast, and Dad can't do that, so it starts there. Dad looks at them and says, 'What can I do? What can I do?' And there's not a lot he *can* do, except earn the money. You can't look at Evelyn and say, 'I know as much about this as you do, dear. I'll go to the doctor with her and give you a break.' Evelyn wouldn't let you do it. And you *wouldn't* know more. And what happens is that somewhere along the line, Mum's ownership of the problem develops a life of its own. She becomes the expert, and he gets left behind. That is why, if anyone ever asks, the first thing I tell them is that under the Bible, your relationship to your husband or wife is the primary commitment. Don't put the kid first. The minute you do that, you're heading for divorce."

Sometimes the pressure between the husband and wife is relieved by appointing a deputy mum, he says. "It's always the oldest daughter. They did a research project some years back and found that an inordinate percentage of eldest daughters, sisters of a child with a disability, never marry."

And the son?

"He has special problems too. He rarely has any responsibility to care for the disabled child, so he usually just drifts off to his room and plays with computers."

And the disabled child?

"Ah, well. She's the center of the universe, an unhealthy focus. If Mum got run over by a truck right now, I really don't know what would happen to Jocelyn. Because she hasn't been trained for independence. She's been trained for dependence. This Sherman tank of a mother just rolls right through the world, and there's Jocelyn sort of riding on the back of the tank."

"Riding very quietly," I say.

"Because for so long, Mum has done all the talking. If Jocelyn ever starts talking, there's a whole Hoover Dam to break." Jackson pauses, eyes fixed on his wall of books. "Most kids start reaching for independence at fourteen. Jocelyn hasn't. And I think when she does, because of the lateness of it, and the amount of investment that Mum has, that Jocelyn will have *huge* guilt in wanting to break the apron strings. And Evelyn will be sitting there with an empty house, with nothing to do."

Which reminds him of when they came back from the LPA convention. Jocelyn came to his office more excited than he'd ever seen her, pulling out a sheaf of snapshots—look how many! Look how active! They even get married and have children! And he realized she'd mulled it over in that inner secret mind of hers and resigned herself to a life without love, and it was funny because he had dedicated his life to the cause of "mainstreaming" not just for political reasons but for God, because the essence of Christianity is that we are all born in His image and then "vandalized by sin," and the distance between Jocelyn Powell and David Jackson is only millimeters compared to their mutual distance from God. Which is why a good Christian values every person, no matter what their age or race or form. And yet there was no escaping the human truth brought home by Jocelyn's photographs, which is that every person with a distinctive disability or even a distinctive gift goes through life feeling like a Martian and feels an overwhelming need to be with other people who are *the same.* But how do you reconcile that tribalism with the dream of unity in Christ?

Jackson sighs and strokes his beard. "I guess what we're all hoping is that, in time, they'll get over the fantasyland of America and find out that dwarfs are sinful people too, and they can hurt you. And you can actually have the same kind of relationship with a fellow dwarf as you would have with anybody else."

With that, he walks me back to the office where Jocelyn is waiting. Writing his Internet address down on a scrap of paper, he asks her, "So you'll e-mail me when you get there? I'll post any messages on the bulletin board like we did before. Because there are a lot of people who care and want to know."

She nods.

"Hope to see you again before too long."

"Yeah," she says, like they both know he's torturing her with these personal intrusions. He gives her a warm grin and she gives him one back, but somehow the grins don't quite seem to connect.

. . .

As we drive away, I tell Jocelyn how much Jackson and the other school officials impressed me. She gives me another one of her noncommittal grunts.

This time, I'm annoyed. "I'm surprised how willing you are to leave people who care so much about you."

"It's true they've done a great deal for me, fund-raising and all."

"I don't mean that. I mean they *care* about you."

"Yes, I know that," she says. "But people move on. It's a school. They expect you to move on, and I want to move on."

"But what is it you think you're going to find in America?"

"To be happy. To find a life."

"What kind of life?"

"Go to university and get a job and have—I don't know, a life."

"Marriage?"

"Yeah."

"Children?"

"Yeah."

"And why can't all that be done here?"

"I don't *know*," she says, laughing at my persistence. Then she stops laughing and tries to nail it down, her voice as flat and analytic as ever. "I keep being met with difficulties here. I suppose in America, I've got hope for the first time."

. . .

That night we go for a good-bye dinner with Jocelyn's best friend, who turns out to be a chatty little Metallica fan with a sleepy eye and a lopsided grin and a teenage penchant for "like" and "y'know." They met in a wheelchair group two and a half years ago and got along great from the very start because Jocelyn likes to listen and Melinda loves to talk, and she knew what it was like to see so many doctors and nurses and rehab counselors who are always telling you what to do and stuff and forget about you the minute they walk out the door, and the friends who want to help you all the time or feel like you're a nuisance.

Melinda's family is the jokey kind, full of quips and digs. At dinner, her dad teases her for being "a little alcoholic." Then Melinda tells a few cripple jokes and teases Jocelyn for being so quiet.

Jocelyn gets her stubborn look. "I have to fight to get a word in edgewise."

"I timed her one night, and she actually talked for six minutes. I think that was a record."

Jocelyn blushes. "You were falling asleep, so I was finally getting a chance to talk!"

Everyone laughs. Jocelyn's bright red and happy, the happiest I've ever seen her.

. . .

From the backseat, as we drive to their house to do the *Today Tonight* interview, Nicholas says he doesn't want to be part of something phony. "I don't think it's right that we represent ourselves as a family."

"We *are* a family," Jocelyn says.

"No, we're not. We're a family that's split into two parts. The kids are a family, but the parents aren't."

When we get to the house, Evelyn is already waiting. The TV reporter is on his way. "I feel like I'm in somebody else's house," she says.

Alecia comes in from her bedroom. "As soon as they get here, I'll call Dad and he'll rush over." Then she walks out again.

When the reporter arrives and the camera crew starts setting up, Evelyn goes into a completely professional mode. You'd have no idea her whole life was falling apart. "Now that she can walk," she tells the reporter, "we want the next step—we want her to have a life." Then David arrives and shakes hands with the reporter, ignoring Evelyn. When it's time to start the interview they take their places on a couch underneath the bookshelves, all four of them crammed elbow to elbow as the reporter prepares them with reminders about cross talk and the boom mike. Then he cues the cameraman and turns to Jocelyn with his first question. "When you came out of that operation and you realized the length of it, and Mum had told you what had gone down, how did you feel?"

"It was pretty typical," says Jocelyn the Withholding.

After a few more tries, he turns to David and asks about the shock of learning that Jocelyn needed a second round of surgery.

"Well, to me, this isn't life-threatening surgery," David says.

When the reporter's finished—no doubt wondering why they are all so strangely stiff today—David follows Evelyn into the kitchen. "Are you taking those X-rays with you to the U.S.? Because I'm going to see Dr. Sillence on the twenty-second, and I'm hoping I can convince them to contribute part of the cost at least."

Evelyn listens with one ear, going through the motions and nothing more. When the camera crew goes outside and shoots Jocelyn romping in the backyard with the family dog, she asks David if he's showing up for the last meeting with their divorce lawyers.

. . .

This morning they gave her electric wheelchair to charity. The push wheelchair goes to Baltimore and then straight into a trash bin, if Jocelyn has anything to say about it. She still has fantasies about using that acetylene torch.

Then we go for a drive down the coast, mostly to get out of Evelyn's hair.

Jocelyn's grumpy because her dad asked her to spend the night with him and she refused, and then Alecia called up to say that Dad was crying because everyone is going to be with Mum tonight and nobody is going to be with him. And Jocelyn still refused. "He's just going to try to make me feel sorry for him," she says.

"You're so tough," I tell her.

As we continue down the coast, she begins to talk about the shock of losing her legs and what a drag it was being dependent on everyone, turning the house upside down so she could navigate it. And then buying the new house and all the arguments her parents had about that, from the size of the kitchen to whether or not Evelyn could have a wood-burning fireplace (inefficient, her father said). Then she switches to Melinda and running wheelchair races at the shopping mall and naturally Melinda went so fast she almost hit people and got detained by the security guard! For the first time since I've known her, Jocelyn is volunteering full sentences, sometimes even two or three sentences at a time.

Then she tells me a secret. "On the Internet, hardly anybody knows I'm a dwarf."

"Why not?"

"Because I haven't told them."

"Why not?"

"You've got to get to know them before you can really spring it on them."

"And how do they react?"

"A few have nicked off, but most of them have stayed."

Later she returns to the subject of her dad. "He comes across so nice, but he's not. If he says, 'I feel sorry for you,' it's not 'I feel sorry for you 'cause you're sad,' it's 'I feel sorry for you 'cause you're going through it with your mum.' " Like the early days of the breakup, when he tried to tell her his side of the story. It was all for him and not for her. "He drains everything out of you. But I don't have much to drain out. I'm barely surviving as it is."

"Really, Jocelyn?"

"Yeah! It takes me a lot to move around every day. And he takes all there is that's in me. And I can't cope. I just fulltaybots."

"You what?"

"I just fulltaybots."

"Fulltaybots?"

"Fall—to—bits!"

She blushes like she did at Melinda's, indignant and happy.

After an hour we stop at a lookout and Jocelyn hobbles along beside me on her canes—using two, she can walk short distances. The scenic marker tells us that this particular bay was discovered in 1876 by three Aussies in a little sailboat called the *Tom Thumb.*

Jocelyn makes no comment. We take pictures.

Back in the car, she gives me her version of the split with her schoolmates. They were pushing her to smoke and drink and said she listened to her parents too much, and then they wanted her to help them with their homework and she refused. "They should've been able to understand it themselves."

"Why do you have to be so tough?"

Her answer is unhesitating and merciless. "I've always had to do things for myself, and I've learned through them. So I didn't feel as though I had to help someone else out of a sticky situation."

. . .

Kitty's in the kitchen making a spaghetti dinner and Evelyn is off packing and fuming—this morning her brother called her a freeloader and she's been in a cold fury all day. Then Nic and Alecia come for their sleepover and Alecia walks right past Jocelyn without saying hello and dogs are barking and toddlers howling and the TV's blaring and—hey everybody, it's time for the *Today Tonight* show! Look, there's Evelyn and Jocelyn pretending to pack their suitcases! Stop laughing so I can hear! Not only are they pretending to pack but they're also using Alecia's room and pretending it's Jocelyn's since Jocelyn's room is empty. And they're packing Alecia's stuff since Jocelyn's stuff is already gone. "That's mine!" Alecia cries. "It's all lies!" Evelyn grabs a teddy bear and hugs it to her chest and then everyone laughs again because what the hell, it's all show biz. You gotta give the public its little jot of pathos. Now the reporter's asking Nicholas if all the sacrifices have been hard. Nicholas puts on a dumb earnest face, answering in a tone of rote piety, "Jocelyn's worth it."

Nicholas groans. "That's the first thing that came out of my mouth!"

Evelyn kisses him. "Those are usually the ones from the heart."

After the show, Evelyn disappears. A little bit later, Kitty has dinner ready. She's one of those harried housewives with a wry, efficient manner. "Where's Evelyn?" she asks.

"Upstairs on the Net," Nicholas answers, rolling his eyes.

"We should have told her the line was out."

At dinner, Evelyn notices that Nicholas is eating just the meat of his garlic bread. "Eat those crusts," she says.

He ignores her.

"I understand pizza crusts, but not bread crusts. Eat 'em."

This time he obeys. But after dinner, he takes me upstairs and shows me all the downloaded *South Park* sounds he's got stashed on the computer, the nasty little voice rasping "I am Cartman! I am Cartman!" He's still trying to impress me with what a dangerous rebel he is. "If your mom knew you were into *South Park*," I say, "she'd have a heart attack."

He laughs. "I am Cartman."

Later I carry some suitcases down to the car, and Evelyn comes outside and demands a hug. "Tomorrow can't come fast enough," she says.

. . .

One year since the LPA convention in Atlanta! One year exactly! In a few hours, another hotel lobby over in that great warm land of America will be filling up with dwarfs and hope! Evelyn rubs the sleep out of her eyes and asks Nicholas to come with her because she has something to say. They go outside, into the backyard, away from the houseful of people, and Evelyn says I love you very much and you've given me a wonderful gift by coming to stay last night and I noticed a change in your behavior and your attitude over the last couple of days and I'm *very* grateful for that. We've always had such a close relationship, always communicated in such a strong and clear way, and no matter how angry we've been there's always been that love underneath. That's why I've been so very saddened by the way you've been reacting this year, she tells him. You need to control your anger. You need to accept the situation. I didn't wake up one morning and think, Well, I'm bored so I think I'll create chaos within my family. I didn't set out to create all this pain and hurt for everybody. But I didn't want to live a lie. I didn't want you and Alecia and Jocelyn to say years from now, "Gosh, those last few years at home were so miserable, when Mum and Dad never spoke to each other." And I don't have a parents' manual that gives me the answers of what to do. I've never done this before, and I hope I never will have to do it again. But I'm not going to justify myself to nosy people like your dad's sister who want to know why, because it's just something that happened through circumstances. I just don't love your dad anymore. I don't want harm to come to him, I hope he finds happiness, but I don't love him anymore and I don't want to stay married to him anymore and these last months have been absolute hell. Nobody rings to see how I am and nobody rings to ask about Jocelyn and everybody believes the lies and the innuendoes and the stories that have been circulating and they don't even have the strength of character to ring me and ask, "Why are you being this terrible bitch? Why are you having an affair?" They've judged and convicted me without even giving me a decent trial. And the reason I went to Texas is because I desperately needed a friend, needed someone to listen to me and give me a hug and say "you're a good person" and it wasn't this wonderful romantic holiday either—I did *not* have an affair, full stop. I got a hug and a shoulder to cry on and by five every afternoon I was locked up in my hotel tower watching TV and ordering Chinese. And if it wasn't a sensible thing to do or a logical thing, I don't know and I don't care. It was an emotional thing and to the day I die, I will *never* regret it.

Halfway through this outpouring, Nicholas crawls over and sits in her lap. He cries. He hugs her.

And stop arguing with your sister, Evelyn says. She loves you and she's desperately trying to keep the three of you together as an intact family unit. And Evelyn goes on about that for a while because she's all wrought up and can't stop talking, there's so much to say—so much that really just comes down to this: *Wherever I am, I will always be your mother.*

Nicholas hugs her again, his eyes wet with tears.

. . .

At the door, Kitty gives Evelyn a Pooh Bear key ring. "It's for all the new keys in your new life," she says.

In the car, Jocelyn and Evelyn don't say much. They look out the windows.

At the airport, David and the kids and Kitty and John and Judy are all waiting. David looks surprisingly relaxed, his eyeglasses dangling from a strap around his neck. And there's Melinda and her mum and Evelyn's mum and dad, who have the sun-baked and fragile look you might expect from deeply religious old Lutherans who have ventured out of the Australian bush to watch their daughter do something unprecedented. The matching white tennis shoes make them look especially harmless.

Evelyn's mum says a stiff hello, then mutters, "David is dead tired." Evelyn ignores it. She's so stiff and tense, she seems to be wearing invisible blinders. Then the ticket lady says there's a problem because they can't go on three-month visas unless they have return tickets, and Evelyn gets scary calm and says I checked this and checked this and I'm *sure* it's right. After three minutes of hell it's cleared up (she was right) and they check the bags and almost check the wheelchair, but then there's an agonizing long stretch of dead time as they wait for an attendant to bring an airport wheelchair and everyone stands around making small talk. Surrounded by family, Evelyn sticks close to me.

On to the food court, where David takes a seat next to Evelyn. "Good," she says. "I wanted to talk to you before we left,"

I get up to give them some privacy. A few minutes later Evelyn approaches me at the coffee bar with a disgusted look on her face. They had a squabble over her divorce settlement, she says. He told her he was going to hold back a few thousand dollars to pay the closing costs on the house, not that he's actually planning to sell the house, and she called him a miserable bastard. "I'm not going to say any of the nice things I had planned for the departure gate," she says. "If he's really in pain, let him suffer!"

She ignores her parents completely.

Then it's time to go. At the gate, David bends down to hug Jocelyn. He tells her he cares for her and he's going to come and see her soon.

"Fine," she answers. "Look after yourself."

Nicholas and Alecia swoop in and hug their dad, then cover the awkward moment by posing behind Jocelyn while John takes a picture. David smiles for the camera as if it were just another vacation snapshot. Then Jocelyn says she wants a shot with just the three kids. Alecia says, "Can't Dad be in it too?" Jocelyn shoots a look over to her mum and makes a scissors motion with her fingers. "He's out far enough to go *kghh*," she says. But nobody notices. Alecia bends down and hugs Jocelyn and tells her that if she ever needs her just to ring or write, and Jocelyn tells her to do the same. Nicholas bends down for a kiss and Jocelyn tells him not to be so mean. Then she goes over to Melinda's wheelchair and for the first time she starts to cry. "Look after yourself," she says. "And look after your mum and study. And study for me too." Now both girls are sobbing. A few feet away Evelyn hugs Alecia and Nicholas and tells Alecia that she talked to Nicholas and he's going to try very hard to control his anger and not to argue with her. And she must work hard at being his big sister and taking on that role of communicating and maybe even discussing the feelings that they're going through as kids because Mum can't help with that and Dad can't help with that. "Maybe you can make the compromise that on the nights that you have rice, if he doesn't want any rice, he'll have the vegetables." And they laugh and Evelyn kisses them both, saying how much she loves them and that she'll *always* be their mother. And if they need her all they have to do is get on the computer or get on the phone and whatever the problem is, they can work it out together.

Then she turns to her mother. "I love you, Mum," she says.

Her mother says she'll pray for her, and Jocelyn too. She promises to check in on David and the kids and help them with any financial support they need. "I wish you well," she says.

Evelyn turns to her father. "I love you, Dad," she says. He nods and stands there in his straw hat and white tennis shoes, looking woeful and stiff.

With that, Evelyn turns to the gate. She takes one step toward America and then stops as if a big invisible hand snagged her. Turning around, she goes up to David and gives him a hug and a kiss. They whisper a few private words. Then David starts crying and Alecia and Nicholas come forward and put their arms around him. Quickly Evelyn turns and pushes Jocelyn toward the gate.

This time she keeps on without turning around and as she passes through the gate she seems so implacable and grim that for a moment she's like a figure out of ancient myth, an Amazon cutting off her own breast so it won't interfere with her bowstring.

And the doors close behind us.

. . .

Near the boarding gate, an elderly man approaches Jocelyn. He says he saw her on TV last night and he wishes her all the best. Jocelyn nods and forces a smile.

A few minutes later, a gray-haired woman comes up to Jocelyn and says exactly the same thing, that she saw her on TV and wishes her all the best. Jocelyn nods and forces another smile.

On the plane, the person in the next seat recognizes Jocelyn, too. "I was wondering if you'd be on my flight," she says. She wishes her all the best.

Jocelyn squeezes out one last smile.

. . .

Once we're in the air, Evelyn orders a Johnny Walker. Then she and Jocelyn start going through all the cards and notes and stuffed animals. David's note tells Jocelyn that he'll always love her and be respectful of her wishes, and Melinda says she's a friend for life and distance won't change that, and Alecia tells Evelyn she'll always be her mum and she's not angry and she's a big girl now. Evelyn wipes her eyes and sips her drink. Then she picks up another letter and her face goes sour. This one attacks her for "taking money from a community which does not appear to be good enough for you" and suggests that she encourage her "new friends in the U.S. to raise the funds instead." It's from an ex-friend who took David's side, Evelyn explains. Putting aside the letters, sipping steadily on her drink, she talks about the innocent early days of marriage when she was so eager to please she'd get up early to put on makeup before David woke up, how worthless she felt then and how she used to tell him he should have married a better wife. And as the children came and she started making decisions, she realized she could make decisions and good ones and she started to stand up to her mother, and then the problems came thick and fast and she began to get impatient with people and maybe that's why—oh, she doesn't know *why* she pushed David away in Baltimore. Maybe it was him. Maybe it was her. Maybe she really was just focusing on Jocelyn. Even now at the gate, she was so mad at him about the settlement that she was planning just to walk away, then at the last minute she couldn't do it. No matter how bad things got between them, they always had a kind of pledge to say a loving good-bye just in case something happened. So she wished him well and told him to take care of himself, and he wished her well and said he knew she would make it.

And now she's going to America. On Saturday, Terry's coming to Atlanta for that visit. She sighs. Her first real date in twenty years. All the changes! "Who would have thought, six months ago, that I would leave my kids? Who would have thought that I would leave my home?"

Quietly, she stares into the back of the seat ahead and says, "I'm looking for something for me now."

And guilt is a wasted emotion. It's pointless to look back and wonder if or why. The important thing is, she saw David this morning and said to herself, I don't love that man. "I don't feel guilty about a thing," she says. And now that she thinks about it the real problem in Baltimore was that David was trying to drain emotional energy away from her when he *knew* that she needed it for Jocelyn. "When we left Baltimore, this is where we were heading."

Enough! "I said to Terry last week, 'Are you expecting any phone calls?' 'No.' 'Are you going to bring your computer?' 'No.' 'Are you expecting any visitors?' 'No, why are you asking all these questions?' 'Because I don't want the phone to ring. I don't want any visitors. And I don't want to see any computers.' "

"No computers?"

"I won't be on the computer all day in America," she promises. "I won't have to be."

. . .

While her mother talks, Jocelyn tries to watch *Blues Brothers 2000* but after a while she climbs out of her seat and stands in the hall, shifting from foot to foot and sulking.

"What's the matter, Jocelyn?" Evelyn asks. "How can I make it better?"

Jocelyn scowls.

"Can I do anything for you?"

"No."

"What's bothering you?"

"Nothing."

A few more hours go by, and Jocelyn doesn't say a word. She sits with her leg straight out, Nike pressed flat against the back of the seat ahead. For six hours now, Evelyn has been talking nonstop and Jocelyn has been sitting there like a squat little sphinx, and it hits me again that Jocelyn is the one driving this, that her silence is squeezing out this vast galaxy of words and maybe even driving her mother's love affair with the perfectly disembodied world of the Internet—that by teaching her mother to distrust the flesh and by alienating her from the world of "normals" who stare and point, Jocelyn has infected her with a kind of sympathetic dwarfism. I remember a passage from Joan Ablon's book where a dwarf's tall mother said she began to feel like a dwarf herself. Later I look it up: "The feelings I had weren't really nice feelings; I'm not proud of those feelings. I felt like *I* was the dwarf. I saw *myself* as being abnormal and stared at for the rest of my life. Those were really awful feelings." And since Evelyn couldn't turn back to the flesh-denying god of her mother and father, she instead turned to the Internet—a vast electronic prosthesis for someone who has learned that the body is not enough.

Then Jocelyn covers her eyes with her hand. She's crying. Evelyn leans over and hugs her and strokes the side of her face and fluffs her hair and keeps up a steady cooing babble, and little by little it comes out: Jocelyn misses Melinda.

"We've been trying to figure out for the last two hours what was bothering you," Evelyn says. "You have to talk about these things."

Jocelyn's whimpers turn to sobs and as I sit there, helpless, I feel a rush of affection for her, for the jaunty way that straight little leg pushes against the seat and how she sits so erect from the rods in her back, leaning her arms on the brace. And when the morning sunlight lights up the little square windows, she cheers up, correcting me with a touch of humorous contempt about my pathetically lame estimate of our arrival time—clearly I know *nothing* about flying to America. I feel a surge of pleasure. This isn't the easy smug perk of "I saw you on TV, I wish you all the best." It's something earned, something true.

After we pass through customs, Evelyn tugs on my sleeve. "I'm in," she says. "I'm in, I'm in!"

She rushes to a phone.

"Terry? Pick up the phone, please. Guess where I am! Terry! Pick up the phone!"

He doesn't, so she tells his answering machine she'll call from Atlanta. Hanging up, she smiles. "I can just see it right now. A black man, a white woman, and a dwarf."

At the gate, I kneel down and give Jocelyn a proper hug good-bye. It's the first time we've touched cheeks. Her skin is so soft.

Eleven

THE WRECKAGE IN AUSTRALIA SENDS ME LOOKING FOR ANSWERS, and one day I drive down to Baltimore to talk to Dr. Kopits. His clinic is a modest suite of offices, identical to every other doctor's office except for the low chairs and stools and pamphlets on dwarfism. When he arrives, he takes me into an examining room and leans against the stainless steel bench and asks me what I'm writing about this time. When I tell him what I saw in Australia, he immediately starts to nod. "This is the great subject," he says. Then he stops, as if caught by the subject himself.

I wait.

After a moment, he continues. "What you are looking into is the abyss. This takes you to the very heart of a human being, to the deepest aspect of the soul." He gives me one of his solemn looks. "Because the thing is, you have to confront yourself." Another pause. He lowers his chin, looking deeper, as if he's trying to peer right into me. "You know?"

And he keeps staring at me in that grave way. I stare back, trying to meet the challenge in his eyes. It's an odd moment of silent communication, almost devotional, as if this were a subject too deep for words. Finally he's satisfied. He gets a chair, sits down, and starts to talk. His grandfather was a distinguished Hungarian orthopedic surgeon and his father was a distinguished Hungarian orthopedic surgeon and he spent his fifteenth birthday helping his father amputate a leg. His father's specialty was osteogenesis imperfecta, a disease that makes bones so brittle they break constantly, leaving its victims so stunted and twisted they are known informally as "pretzel people." But his father didn't just study them, he brought them home to dinner and made friends with them, a memory that must have lingered in Kopits's brain as he went into medicine himself, flirting with dermatology and gynecology before finally settling on pediatric orthopedics. Children's bones were still growing, not quite as complex as the pretzel people but still a challenge. Then came the day his life was transformed. The date is fixed in his memory—January 28, 1968. He was thirty-two years old. A father came running into the Johns Hopkins emergency room carrying his three-year-old son in his arms, the boy's arms and legs flopping helplessly. Kopits ordered X-rays and learned from hospital records that the boy had come in recently with complaints of breathing trouble. The pediatrician who examined the boy for the breathing trouble couldn't find anything wrong. Could the two incidents be related? Lungs to spine? Then the X-rays came back, revealing a dislocation of two neck vertebrae. Kopits knew that if those floating vertebrae pinched the spine, it could cause permanent paralysis or even death. Later he learned the father had slapped the kid that morning and was tearing himself up with guilt, but no slap caused this—the bones had *grown* dislocated, grown so far askew that almost any bump could have popped them apart. He'd never seen anything like it and had no idea what could have caused it, but he had to act fast. So he made a quick decision and fused the boy's neck bones, threw himself into the task of solving the mystery. He learned that the boy's father suffered from asthma and took great pride in having overcome his limitations. He wanted his son to be tough too, a fighter who would push his way through the difficulties of life. But the son was weak. He also seemed to have breathing trouble so the dad thought it was just asthma like he had and kept pushing him to be strong, to try to keep up with the other kids no matter how he felt. But the boy had other problems too. His fingers were so clumsy that he had trouble unbuttoning his overalls. When he had to pee, he seemed to get even clumsier. The father was only twenty or twenty-one and didn't deal with this very well. And that January morning when the kid wet his overalls again, the dad slapped him and the boy collapsed. He had no way to know that his son's weak hands were a symptom of compression of the spinal cord and his breathing trouble the result of spinal pressure on his lungs. And there were other symptoms not yet visible to the unassisted eye, the shortened arms and legs, the slightly enlarged head. Once Kopits put it all together, the diagnosis was obvious and yet—and this is the thing that moved him and changed his life—the boy had almost died from the lack of one.

Kopits plunged into a study of dwarfism. He learned that dwarfs had many orthopedic problems, that gravity punished them and their own body weight punished them and that many couldn't reach below their knees or touch the tops of their heads or even wipe their bottoms. Each passing year took a toll, often putting them on crutches and into wheelchairs, but in most other aspects they were normal human beings and maybe that was the most disturbing thing of all, the reason why so many people assumed that dwarfs were mentally retarded. Because it was easier to think they were shielded by stupidity than to think they were fully aware inside those difficult bodies. And yet so many of them seemed hopeful and happy, which was a mystery and a miracle. And nobody was working on this, not in the field of orthopedics anyway. Here was an area of medicine with no established procedures, no creative solutions, and nobody really wanted to do the work because it was so complex and seemed to be a professional dead end—there weren't that many dwarfs out there, and they weren't exactly beating down the door looking for treatment. It was the perfect way to follow his father, right up the steepest face of the mountain. So he kept doing research and soon he began to get referrals from other doctors, and the more compressed necks and bowed legs he saw, the more he realized that he couldn't fix the necks without addressing the problems caused by the bowed legs, and he couldn't fix the legs without treating the hips. He moved from pediatric orthopedics to the Moore Clinic, where Dr. Victor McKusick's push to study the genetics of dwarfism was bringing a steady stream of patients. There he was, at the dawn of a new era!

But he saw tragedies, human tragedies, heart-wrenching tragedies. That first patient, the little paralyzed boy, ended up out in California in a home for the mentally disabled. The parents couldn't handle it and got divorced. When Kopits found out, he flew to California to see what he could do, and the boy still had his picture, pulled it out of his pocket the minute he walked in the room. Kopits got him released, got him into a normal school, but that was still just a glimpse into the abyss. Because one thing Kopits has learned through bitter experience is that the secret whispers of the fearful human heart and the dark ancestral belief that there is some kind of justice in this world are powerful and insidious and can destroy you. And no matter how often he tells parents their dwarf child is a random mutation falling from the heavens and nothing they did or their parents or their grandparents did has anything to do with it, people confronted with a deformed child face the supreme human challenge. And the horrible thing is they know it—they *know* in their hearts that their response is going to be the yardstick that measures their humanity. And this supreme challenge is thrown at people who are completely unprepared for it, and they are bound together or ripped apart and show heroism or cowardice. And even the heroism can rip people apart, as well as hospitals and time and distance and all the little fears and resentments that come between people who inevitably move at different speeds to confront the challenge. And just to make it all the more maddening, it is a day-to-day response, not a decision made once and forgotten, because sometimes they love the child and sometimes they hate the child and sometimes they just want it to go away. And this is the important thing, he says with emphasis: *The more guilty they feel, the more they reject the child.* This is what I read about in the textbooks, in the distanced language of academics, but there's no distance in the way Kopits tells it. For him it's a living tragedy, a personal tragedy. Because the problem starts when you take it personally, and we *all* take it personally. Even doctors. And even journalists, he adds, looking at me with that penetrating solemnity. This is what he has learned after thirty years in the little world: *Dwarfs can bring you down to the basics of what you are.*

He's on a roll now, not waiting for questions. "I've been waiting a lifetime to tell someone these things." Say for example the mother sees this physical defect as something lacking in her child, her creation, and the mother usually does because it's hard not to. So maybe she shuts off. That's one response, not uncommon. Or maybe she throws herself into finishing the job, trying to make the child complete by heroic lavishments of love and nourishing, and soon she quits her job and starts haunting hospitals and becomes more and more convinced that she is the key to all progress—with those short arms, the kid can't even wipe his own bottom! So Mom does it and doing it becomes so meaningful and important and fulfilling and exhausting that she barely has time to talk to her husband. She gets testy, brittle. This is her job, her mission, her calling. She is the scholar of the book of the disabled child. And then the kid gets to be ten or twelve and Mom is still wiping his bottom and he goes to occupational therapy to learn how to do it himself and then *she* needs therapy, she's developed a "reverse dependency," and the bathroom breakthrough threatens her identity as Saint Mom. And consider the fathers. Lost in the background, paying the bills, neglected by all but in some ways suffering the most. Because a father naturally sees the child as an expression of himself and wants the kid to go him one better, to be the ballplayer or artist or business success he never was. But what kind of big shot can a little man be? What social stature for the short-statured? So the father feels helpless, a failure genetically and a failure as a man, and there's Mother Courage off healing the kid and completing the kid and everyone admires her. And he can't even admit how he feels because then he's a cruel bastard in thrall to his "male ego." So eventually he tunes out and "leaves his face behind," as Kopits puts it. He has seen it many times. Or he starts drinking or walks out or maybe even kills himself. Kopits has seen that too, twice, the poor failed man hanging from a roof beam or driving into a wall. And once a father killed the child. Tragedies. Human tragedies. A glimpse into the abyss. So what could he do but petition Hopkins for more research money and, when they wouldn't even give him an assistant, start the Little People's Research Fund? He needed his own power base to get this work done. And let's please go off the record to discuss the resentment and bad blood that still linger, the gossip from other doctors and even some members of the Little World. Because when you do this kind of work, you see real results, you make people *rise from their wheelchairs and walk.* And when the media started coming around, why not spread the word? And yes he did agree to that "saintly doctor to the little people" article in *Parade* and the one in *Reader's Digest,* and accepted the Baltimore's Best award. But why would any reasonable person be jealous? Money had to be found, research done, hips, knees and spines studied. And didn't he have to streamline his life and push away everything that wasn't important? Isn't that the only way to accomplish anything, to be absolutely focused? Did it really matter if he found it difficult to make pleasant collegial chitchat with certain local doctors or the damned Italians about the wretched practice of leg-lengthening, turning an obsession with human beauty into a rationalization for torture? Breaking bones and stretching them with armature just for a few inches? Fine if it's done to correct an imbalance in length, but for cosmetic reasons it is simply barbaric. So that's why it finally made sense to move across town and start his own skeletal dysplasia clinic here at St. Joseph's, taking the research fund with him and hiring a staff of nurses and an operating team and Sister Celeste and starting the Pierre House and all the rest. And with all this on his hands plus the research for his book and spending two and three and even four hours with each patient, could he really be expected to go home every night at six? Could he find time for his own children and his wife? Could he keep his own marriage together? "When you rip things apart," Kopits says, "you are coming down to your basics, to the basics of the human being in a way most people have no idea—they have no idea what their basics are, because they have never had to confront their inner self in such a way. I myself have—I see so much tragedy, on such a daily basis, and for so long, that I cannot really—I cannot look at all I know. I must somehow seal it from myself."

. . .

Talking to Kopits forces to the surface some things I felt but couldn't quite put into words. All year, I'd been tiptoeing up the edge of this story and quickly backing away, pushed and pulled by a sense of guilt and responsibility and affection and friendship and fascination on various levels, as well as a steady stream of uncomfortable glimpses into my own character. Maybe even some kind of religious impulse mixed in there somewhere. But the truth is, I was hoping the story of Evelyn and Jocelyn was over, the flight to America a nice, upbeat, into-the-new-world ending. Now I see that I'm going to have to go through surgery with them, right into the operating room if I can.

Michael and Meredith are another sore spot. Since that uncomfortable lunch in New York, I've had dinner with Meredith a couple of times and gotten back on friendly terms, but Michael has continued to be punishingly silent. Once or twice on the phone, he admitted a tendency to paranoia and said he might be overreacting, but he wanted to see how things went at the next LPA convention before making any commitments—if things have calmed down, then maybe he would take a chance on a few more interviews. So I wait for the convention to end and wait a few more weeks and then call Meredith to see how things went. No response. I call Michael. No response. A week later I call them both again, and this time Meredith gets back to me and says the reason they aren't responding is because the minute they walked into the hotel, somebody marched right up and asked, "Are you the ones who said we had big heads?" And it went on like that *for five days.* She thought he was being paranoid too, until she saw it for herself. It was awful. And people still blamed Michael for all my questionable lines, even more than they did on the listserv. The letters I'd written to the listserv and the LPA newsletter didn't do any good at all, maybe even made things worse. And it doesn't matter that they were misreading, doesn't matter that it's a year later. They were still crazy over that line about him being the best-looking dwarf. In fact, they insisted it was his line and I'd just written it down. Plus, some women gave Meredith a really bad time about waltzing into her first convention and *taking center stage* in a national magazine. And a group of Australians cornered her in an elevator and even though she told them she didn't say any of that insensitive stuff about big heads and told them she didn't like it either—and it really *was* unacceptable, John—she was trapped. It was scary. They just *attacked* her. And the more I try to plead my case, the more stiff and cold she gets. "Michael feels burned," she says. "He's learned a lesson."

"What about you?"

"I love Michael," she says.

Then Andrea calls to ask if I'm not talking to her anymore.

"You're the one who hung up on me," I say.

"I didn't think of it as a hang-up."

"As I recall, you said, 'Find someone else to raise your consciousness,' and hung up the phone. When someone issues an ultimatum and hangs up the phone, I generally take that as a hang-up."

I am giving her more shit than she deserves, probably. But to hell with it—she's short, and she's there. Except it turns out to be another big mistake because it takes us right back to the conversation about me rejecting a basic part of her and it goes on and on and on and on until I explode. "Look, Andrea—when I use the word 'wrong,' I *clearly* mean wrong re the norm. I'm tired of getting shit for this."

"Then just say different. Different is different than wrong."

"But my daughter's buck teeth look wrong to me! That's why I'm spending thousands of dollars for orthodontics!"

. . .

Once again, Andrea's anger sends me to the library, where I start digging out books with titles like *Body Images: Development, Deviance and Change.* The books tell me that philosophers and theologians have been arguing and thinking about the split between the mind and the body for thousands of years. Although there have always been "materialists" who insisted on the fleshiness of life and those Christians mentioned earlier who believed our bodies were made in God's image, for a long time the argument seems to have been dominated by idealists like Plato and St. Thomas Aquinas who believed that body was of the earth (or lower) and the spirit was up in the clouds with God. That's what caused the scandal when Freud began to argue that the mind was all tied up in the body—castration anxiety and penis envy and all that. Not only was he making all these embarrassing sexual references, he was pulling against the basic religious impulse toward idealism, tugging us down from God. And Freud's followers took his ideas about the body into a vast number of different directions. Carl Jung developed theories about the body as "protective enclosure," the biological model for primal religious archetypes like the Garden of Eden. Wilhelm Reich introduced the concept of "body armor" and suggested a vast revolutionary program for casting it off. Erik Erickson argued that women were less armored than men, an idea that would echo decades later in the school of "difference" feminism. Paul Federn focused on the boundaries between the individual body and the world with probing questions about why the body becomes so permeable in dreams and why weak boundaries make the world seem so "strange and depersonalized." But for our purposes, the most important thinker was a man named Paul Schilder. He developed the influential concept of body image or "body schema," arguing that every emotion we feel is reflected in our flesh. When we hate our bodies, we contract into ourselves, hunching our shoulders and presenting hard outlines to the world. When we love our bodies, we open our arms and invite the world inside. And when our bodies are not enough in themselves, we experiment with clothes and masks and jewelry to "expand, contract, disfigure, or emphasize the body-image and its particular parts." More directly to the point, Schilder argued that a sense of one's body as strange and alien is a part of every neurosis and "beyond doubt" due to "withdrawal of libido from the body image." In other words, when you don't feel attractive, you don't feel sexy—and that makes you crazy.

In Schilder's wake, thousands upon thousands of papers and studies and books have been written about body schema and body image, and thousands more about deformity and its psychological impacts. Skimming this material, I find the connection between the body and the mind everywhere affirmed: "A unique closeness exists between one's body and one's identity," one scientist writes. "Although objective physical attractiveness is only modestly related to personality traits, subjective physical attractiveness is strongly related to them. People who view their bodies unfavorably tend to suffer from a variety of problems. . . ." Another points out that scientists who study higher primates identify the onset of self-awareness at the point when the apes recognize themselves in a mirror. A whole vast literature explores the researches of neurologists who have studied brain damage and discovered exactly how hardwired our body-mind unity is—patients with brain damage sometimes can't tell right from left, or claim that part of their body is being "used" by someone else, or deny the existence of certain parts of their bodies or insist that they've suddenly grown second heads or third arms or horns invisible to the eye. Some scientists argue that body schema is the basic frame of reference we have for interpreting all our experiences, that "left wing" and "right wing" are more than handy political metaphors but an inescapable (if somewhat reductive) way of defining the world.

All this makes me feel much better: I'm right, Andrea's wrong.

And better still when I read this: "Body awareness remains permanently a major issue for the disfigured and profoundly shapes the construction of their self-concept. They must constantly work to maintain a sense of self-esteem devoid of reliance upon beauty."

And this, summing up a survey of nearly thirty thousand people that showed that 90 percent of people who reported feeling good about their bodies also reported feeling good about life: "Obviously, if one dislikes the body one 'lives in,' it's difficult to be satisfied with 'the self who lives there.' "

The problem is, it's hard to stop reading. There's something weirdly compelling about this material, as compelling in its way as the sight of a dwarf. I learn that the word "ugly" comes from "uglike," an Old Middle English word that meant "fearful, dreadful, vile," that it was common in olden days to disfigure criminals by scarring or wounding them (as some countries still do), thus rendering them "ritually polluted," and that some of this ancient stigma sticks to scars and wounds and other disfigurements today. That scientists who study people who have been disfigured in fires or accidents use the phrase "spoiled identity" to describe how these people come to feel about themselves. That people who are closer to normal try much harder to close the gap than people who are further away, and are more tortured by the lesser distance. That studies have confirmed Schilder's insight into body image and libido by showing how people who feel unattractive tend to think they're not "love worthy." That one disfigured woman said she had begun to think of herself as a *thing*, a feeling that oddly echoes the objectification beautiful women and dwarfs both disdain. That ugly and disfigured people tend to develop what psychologists call "avoidant personality disorder." And I also see, over and over, doctors and dentists writing about the conflict between what they seem to consider proper medical treatment and the "cosmetic" concerns of their patients—the same word Dr. Kopits used when he was talking about leg-lengthening. And they use the same strangely dismissive tone. I say strangely because this latest bout of research has brought me face-to-face with the mirror image of the beauty research I did before, and it is a dark mirror indeed: studies showing that mothers of ugly or disfigured newborns neglect their children, that ugly children get punished more, that ugly adults get poorer work evaluations, that "unattractive people are more apt to encounter social environments that range from nonresponsive to rejecting and that discourage the development of social skills and a favorable self-concept." In one study, people who listened to tapes of ugly and beautiful people found the ugly people more insecure *even without being told they were ugly*, which proves that the unfavorable self-concept isn't just a loose generalization or projection of guilt but a demonstrable fact. In a study of schizophrenics, another scientist speculated that not only do schizophrenics tend to be less attractive on average, but that ugliness contributes to schizophrenia. "Physical unattractiveness," he concluded, "represents a risk factor in the development of psychopathology in general."

Studies of the disabled echo and amplify these conclusions. One study found that children with physical disabilities have three times the average chance of developing mental illness; another found among the disabled rates of depression and anxiety two to four times higher than normal. And no wonder, since they are often born prematurely and stay in hospitals for months, causing "attachment problems." And since they tend to be more irritable and immature, which causes more attachment difficulty, a problem that ripples down the years. Another study showed that people with cleft palates get married at half the rate of their normal siblings and also earn less—and live closer to home and have fewer close friendships and report more social anxiety.

But the exchange between the mind and the body isn't always simple or linear. Just as people who are nearly normal tend to be more tormented by their difference than those who could never dream of "passing," people who *think* they are defective even though they're actually quite normal are often much worse off psychologically than people who are defective but see their defect without drama. In one of the most interesting studies of this, researchers had makeup artists put a scar on the face of women and then told the women to go talk to a stranger. But just before each woman went to meet the stranger, the researcher said he had to adjust the makeup and surreptitiously wiped off the scar. "Relative to the control group, those women who *believed* that they had a disfigurement reported more discomfort and alterations of gazing behavior *in the partner* during the interaction. Thus, our body images can alter our information processing and cause us to 'see' what we expect to see."

Complexities of a more personal kind crop up when I start reading about the barrier concept, which was developed in the late 1950s by a pair of social scientists named Seymour Fischer and Sidney Cleveland. After years of study, they concluded that people who have high barriers often "continue to operate relatively satisfactorily at a psychological level" even with a "highly defective body." Often these individuals are unusually ambitious and driven, and wisely so; as the purpose in their lives increases, the importance of their bodies tends to decrease. But "permeable barriers" are definitely a problem. Caused by "overly intrusive" parents who are so physically or psychologically dominating that they cause "a blending of parent and child in the child's mind," or by parents who go the opposite way and don't stroke and caress or otherwise respond to their kids at all, permeable barriers can induce a variety of reactions from transvestitism to exhibitionism to anorexia. As another scientist puts it, "many behaviors regarded as impulsive, addictive, or unrelenting" are attempts to "evoke or establish boundaries." Often these people express a desire to escape the body or shed the skin. When they draw bodies, they tend to blur the outlines and make distorted and overly large shapes. Often they have learned to "organize experiences around pain and illness in order to obtain attention and affection" and are so guided by the responses of others that they "may not even know or have developed true selves." Though they are often very successful and driven people, they also talk of having "a vague sense of incompleteness or emptiness, a feeling that something is missing."

All this reminds me, of course, of the other boundary issues that have come up in this last year, from the way dwarf children tend to draw "firm boundaries between themselves and the outside world" (the first time I heard the phrase) to Andrea telling me how "wonderful and scary" it was when I challenged her boundaries. And Leslie Fiedler's line about the way "human monsters" threaten the boundaries we draw around ourselves by reminding us of a "sacred unity" we somehow learned to divide into pallid but manageable chunks. Which makes it perfectly natural and appropriate that every so often, as I page through these books, I'll read something that crosses my own boundaries, causing an uncomfortable twinge of recognition. One jumps out of the pages of a book called *The Psychological and Social Impacts of Disability*, an almost offhand reference to common emotions inspired by the sight of a deformed person—*existential anxiety* and *aesthetic anxiety*. Oh? Reading on, I learn that existential anxiety consists of "projected fears and apprehensions about human mortality and about the prospect that physical impairments could interfere with a preferred lifestyle." Aesthetic anxiety is caused by "the enormous cultural social importance of physical appearance [and] also on the common proclivity to seek social acceptance by striving for conformity with perceived similarities among human beings and by evading contact with those who are physically different." Another day I come across a reference to "high self-monitors," people who are "especially sensitive to impression management" and spend a lot of time on things like makeup and clothes and working out in gyms and (perhaps) even reading books on body image. Apparently these sad individuals are not only affected by their own looks but are particularly "responsive to the physical appearances of others" and have (as yet another scholar tells me) a tendency to "compulsively pursue physical attractiveness and social affirmation and acceptance to supply vital narcissistic needs, to counter an inner emptiness, and to insure against expected abandonment."

And finally there's this, from a remarkably straightforward little book called *Changing Faces: The Challenge of Facial Disfigurement*. "My disfigurement is so very obvious that curiosity is inevitable. It is ironic that often the people who are the most curious have many problems of their own, though less conspicuous ones. They often wish someone would ask them leading questions, so they can offload their problems. Their curiosity about my disfigurement turns out to be their cry for help."

. . .

Jennifer Arnold lives in Baltimore too, in a tall apartment building just a few blocks from Dr. Kopits's clinic. I haven't seen her since the night of the LPA ball, when she and Martha Holland talked so movingly about the pain of being three-foot women in a hotel filled with four-foot men. So one night I go over to have dinner with her. As she cooks the meal, climbing up and down a stool to reach the stove, she talks about her life as a medical student at Johns Hopkins. Most people are very helpful and open. There was a time when she was learning to deliver babies and a doctor ordered her to stop. He seemed angry that she was even trying. So she got to deliver only one baby. But the patients don't seem prejudiced at all, almost as if they think she has some special connection with them. Once again Jennifer strikes me as formidably intelligent and almost heroically direct. She tells me about the years she spent getting surgery from Dr. Kopits, a total of twenty-two separate operations, her mother sleeping by her bedside. That's probably the main thing that got her interested in medicine as a career. She's been working so hard at it, she's had almost no private life. She's just starting to date now, a dwarf man she met through the LPA. In fact, just a few days ago, she kissed him for the first time.

A few minutes later three of her classmates come to join us for dinner, all tall and strikingly attractive women. As they talk about school and families, the subject of Jennifer's date comes up. They ask gently teasing questions about how it went, and suddenly I find myself blurting out, "She kissed him." Immediately I regret it. One of her friends gives me a sidelong glance and says, "You must really be a good reporter. Jennifer never talks about that sort of thing."

I'm so ashamed. Months later, years later, I think of this and feel the same yellow wash of shame. Why did I blurt that out? What the hell is wrong with me?

. . .

Meredith calls again, says she tried talking to Michael but it's just not happening. The Little World is too important to him, the conventions and the sports and everything, and he doesn't want any more exposure. Never mind if that's giving power to a bunch of gossiping idiots. "I walk down the street and someone calls me a midget—yes, they're idiots, but it still hurts."

I plead my case once again, telling her my latest thoughts about taking center stage and the need to be seen. For a moment, temped by the promise of being understood, she seems on the verge of yielding.

Then she suddenly firms up again. "It's easier for you to hide. You can put on a baseball hat and go to the store and you won't be different than another John. Anywhere I go, I'm recognized."

. . .

Since she came to stay in my house three days ago, Evelyn's been getting up at dawn every morning to send e-missives to a tiny mining town on the desolate west coast of western Australia. After another three months living out of suitcases in yet another small room (they've been staying with an LPA family in Atlanta), after things went sour with Terry and Jocelyn finally blew her top and the money got tight, Evelyn started spending more and more time e-chatting with a fifty-year-old Australian miner named Dave. Talk of their pasts and dreams turned into a dozen roses and then plans for him to fly all the way to America just to see her. Was it love? She had to find out. And she didn't want to be the princess in the tower this time and she needed a break, so she flew to Chicago to meet him. Two minutes after he got off the plane, they both knew. They flew to Florida together and took Jocelyn to Disney World, and Dave was so attentive and fatherly it made Jocelyn cry. Two weeks later, on his way home, he called Evelyn from his layover in Los Angeles and asked her to marry him.

Now they "talk" for five hours at a stretch, eight to ten hours a day. When the Internet's not enough, they talk on the phone. Last month, Evelyn's phone bill was over a thousand dollars.

And Jocelyn has been unusually quiet, answering questions in monosyllables or less. "Tell me about Dave," I say.

She shrugs.

"What did you think of him?"

"Nice guy."

Evelyn looks up from the keyboard. "Talk about the night in the Econolodge when you got really upset."

"I've never been there," Jocelyn lies.

"You were moody."

"Yeah."

"Why were you moody?"

"I don't know."

Later Evelyn tells me that in the Econolodge, Jocelyn started crying even worse than on the plane and said she had no friends and she was lonely and sad and her dad had deserted her. Now she's bottling it up again, nervous about the upcoming operation, just two weeks away.

The next day when I get home from work, Evelyn holds out her hand and wiggles her fingers. On the third finger of her left hand, she's wearing a gold band embedded with tiny chips of diamond. "It's a funny little thing," she says, "but it's just until we get a real one."

"Well, at least you won't get hit on in bars."

She laughs. "Like I've ever been in a bar!"

She's planning a June wedding. In Australia. Dave's got only a few more years at the mines before his pension gets vested, so it's good-bye American Dream and hello red dirt and wallabies. Once she made up her mind, she sat down with Jocelyn. "I told her, 'This is what I'm thinking about,' and she said, 'Oh, I thought you were gonna stay here with me,' and I said, 'Well, that was my original intention, but now there is somebody else in the picture, now I have to take two people into consideration.' "

Jocelyn will continue in Atlanta alone. A mutual decision, Evelyn says.

It's shocking. So much has changed, and so fast. I remind myself of all the pressure Evelyn has been under, of all she's lost in such a short time—her family, her home, her country. This is what Dr. Kopits meant about rising to the ultimate challenge, about coming down to the basics of what you are. "Maybe it would be good for Jocelyn to get on her own a bit," I say.

"She's already more independent. Did you notice?"

When I drive them to catch the train to the hospital, we talk about it again, this time with Jocelyn present. "It's something I have to do, and it's something that she has to do," Evelyn says. "It's come to the time for the bird to leave its nest."

I ask Jocelyn how she feels about that. Evelyn turns to see her response. "It won't really bother me," Jocelyn shrugs. "We'll still see each other, and we'll keep up our contacts."

. . .

Andrea calls on my birthday. "How old are you?" she asks.

I tell her. "You *are* old," she says.

"You're short, so fuck you."

"But I don't feel bad about being short," she says.

We've been talking again for about a month now. She even gave in and said I could use her in this book if I promised not to use her real name. Now she throws out a provocative new twist—she's taking an essay-writing class at a local university, and I'm her subject. It's my turn to be poked and prodded and exposed.

"So send it to me," I say.

"No way."

"Then why'd you tell me about it?"

"No reason. Just making conversation."

"What a tease you are."

Then she really surprises me. "I told my father you were writing about me and I wouldn't let you use my name," she says in that drawling voice of hers, so rich with irony.

"What did he say?"

"He said, 'You won't get anywhere in life that way.' "

"That's weird. What did he mean by that?"

"I asked him. And he said, 'You don't want people to know you.' "

I remember what she told me at the LPA convention, that her grandmother suggested she get a job as a typist because then nobody would have to see her. I remind her of this and suggest that maybe she took Granny's advice to heart, that what she really wants is a life where no one can see her.

"Fuck you," she says.

"Your father's right. You have to stand up and be counted."

"It's not that. I worry that I'm performing."

That word again. And because of the way it came up, I think I know what she's really talking about underneath it all: performing for Daddy. That's true of almost everyone, but with Andrea and me, it's probably our biggest connection, the point where we overlap the most. Except with Andrea, you have to mix in the whole gawking world, a nightmarish amplification of the paternal gaze. No wonder she revolts.

On the other hand, I have learned in this last year something she doesn't yet know—the unfillable emptiness when you're still up there on the stage but Daddy has finally left the theater.

"I think the honest thing to do is just admit you want the attention," I tell her. "As much as you hate being the center of attention, you love it."

"Why do you say that?"

"You're the one who keeps coming back to fight with me. It's not like you don't want attention. So stop fighting it. Send me your damn story."

"Maybe I'll send you an edited version."

"You're too paranoid. You'll cut out all the good stuff."

"I'm still uncomfortable with you being a writer."

"Come on, send it to me. Maybe it will improve my character."

"Nothing will do that," she says.

Twelve

THE NIGHT BEFORE JOCELYN'S FIRST LEG SURGERY, I arrive at the Pierre House to find Evelyn typing away on the house computer. She says hello and lowers her eyes back down to the screen.

I sit down with Jocelyn. "Are you nervous?"

She shrugs.

Evelyn looks up from the computer. "Her stomach was going bongos this morning."

I don't know what's going on with Evelyn lately. I've barely talked to her in the two weeks since she left my house, and haven't seen her "buddy" icon lit up on the menu of the ICQ chat program she uses so compulsively. I may have offended her by asking too many skeptical questions about Dave. Or maybe it was the raised eyebrows when she told me about that thousand-dollar phone bill. So this time I decide to follow Jocelyn's example and leave her alone. Jocelyn and I order pizza, then fence our way through a few questions about life in Pierre House. Yes, there's a dwarf boy her age staying here. No, she won't tell me if he's cute. No, she won't tell me if they've had any conversations. In fact, she'd prefer not to discuss him at all, thank you. We end up watching *Caroline in the City* on the living room TV, and Jocelyn tells me that it's one of those plots where the heroine and her friend love each other but can't admit it. "Have they ever kissed?" I ask.

"Yeah," she says. "A couple of times."

"Have they slept together?"

That gets Evelyn's attention. She looks up from the computer and frowns. "Hey, she's only seventeen."

At ten, Jocelyn goes to bed. I sit with Evelyn for a while and try to make conversation, but she doesn't stop typing for a minute. It's irritating. I'm sitting right here, a flesh-and-blood friend, but she'd rather write to some fantasy man halfway across the world. No wonder David got so frustrated.

When I go to bed a half hour later, she's still typing away.

Upstairs, there are four or five rooms with old lumpy beds. All the light switches are rigged with Lucite straps that hang down to dwarf level and the bathroom has two mirrors, one for tall people and one for dwarfs. Sitting on the steps as I say good night to my kids on the hall phone, my eyes drift over a small bulletin board overflowing with newspaper clips. There's a story about a dwarf wedding, a dwarf real estate agent, a dwarf athlete, all described in the same happytalk disability boilerplate about big hearts and ordinary lives—and look, here's one about Martha Holland. It's eight years old, all yellowed and curled at the edges, with a big picture of Martha and the boyfriend she mentioned, the one who died from a heart disease. He's tall and skinny and wears big glasses, a sensitive-looking kid. The story says that nothing has stopped her, not dwarfism or all her operations, that she's one of forty-four national winners of the *Yes I Can* award from the Foundation for Exceptional Children, an award given to "children who have overcome their disabilities." And her boyfriend loves her for her soul and doesn't care about appearances. And it's all just bullshit, evil happytalk bullshit that just drives the darkness deeper so the normals can feel better and the dwarfs can enjoy the momentary relief of a pretty lie. Anyway, that's what I feel in the moment, sitting there on the steps listening to the computer keys clacking away downstairs.

At five, Evelyn wakes me up. She's wearing a Mickey Mouse T-shirt and a Winnie the Pooh watch. I wash my face and make coffee in the dark and then go back upstairs to find Jocelyn sitting next to the bed talking to Dave, her back so rigid from the rods that she looks like a doll forgotten on the floor. Five minutes later we start the walk across the parking lot to the hospital, Jocelyn's last walk for five months. The sky is still dark and cool, and the predawn silence feels slightly clandestine, a secret moment in a secret life. Every fifty feet, we stop so Jocelyn can lock her canes against her hips and rest.

"I haven't seen you on ICQ lately," I say to Evelyn.

"I went invisible."

As we walk into the waiting room, Evelyn grabs a prayer slip from a small box on the counter. "Father," it begins, "as I face this operation, I come to You with my fears and misgivings and ask You to put into my heart the needed courage to face the day with unwavering confidence in Your goodness and protection." Then we sit and chat and watch the obscenely chirpy morning newsreaders on the TV bolted to the wall. An hour later we go into the prep room and a little after that Dr. Kopits comes in wearing a tie with koala bears and kangaroos on it. "You're wearing our tie!" Evelyn says.

"Of course." Kopits gives them both hugs and then leaves for the operating room. Before long Evelyn and Jocelyn go too. I was supposed to tag along, but there's a last-minute problem with the anesthesiologist, who doesn't seem to like reporters. Half an hour later, Evelyn comes out and heaves a big sigh. "I need a hug," she says.

The waiting begins. Breakfast kills some time, then it's up to Jocelyn's room, where the stuffed bears on the dresser and get-well cards stuck to the bulletin board and the Piglet pillowcase on Evelyn's bed just make everything seem more lonely and sad. "It's hard to believe a whole year has passed since the decompression," Evelyn says, leaning back on Piglet and drifting off into memories of the past. Then she gives herself a shake, gets out her laptop, connects with Dave in Australia. The hours start to drip away, slow and steady as an IV. At 11:30 a nurse calls Evelyn from the OR and tells her that Kopits is still "exposing." Fifteen minutes later, she calls again and a huge grin splits Evelyn's face. "Yes! Yes! Yes! Tell him God is wonderful!" She gives me the news: Jocelyn's hips won't need surgery! All by themselves, pushed by her new back, they rotated into position! She types the hot bulletin into the laptop and then unplugs the modem and plugs in the phone. Even though it's way past his bedtime, Dave wants to share the happy moment voice to voice. "He makes out to be this big mean man, but he's such a softy," Evelyn says.

Two floors down, Kopits is cutting through Jocelyn's bones with his oscillating saw.

When she gets off the phone, Evelyn washes her hair and takes a nap. The minute she wakes up, she pulls out her laptop again and starts checking in with her ICQ buddies around the world. A guy named Matt comes on-line from Holland to ask if there are any results and if she's sad, and Evelyn smiles at the screen. "I may be sitting here alone, but all my friends come to visit me."

At 5:00 P.M., a nurse calls from the operating room to say that both legs are done. Then Evelyn squeals in excitement—Nicholas just logged on to ask how the operation went. Perfect timing! "Is everyone there," she asks.

"No, it's just me."

She details the great surgery news and the response—a comment on the positive financial implications—makes her frown. "Is this Nicholas or David?" she types.

The answer comes back: "David."

"Good morning, shithead," she says.

Soon, David turns the keyboard over to Nicholas. "Hey Mum," he types.

"Hey handsome. Did Dad tell you about Joc's surgery?"

"No he had to get ready because we're leaving in like five minutes."

"Well, she did not need to have her pelvis operated on. This is wonderful news for her. It means less surgeries and less pain obviously. She will be so pleased to hear this news. And I am VERY EXCITED about it. I have been waiting for you guys to come on-line so I could tell you."

"Okay. Okay thanks."

Evelyn frowns, nonplussed. "You are welcome," she types.

At that point, the ICQ chirps at her. Dave's back on-line. She switches screens and her message appears: "I love you with all my heart."

"Hang on, Dave!" Evelyn says out loud, typing furiously. "I'm talking to the kids. I'll talk to you when they go to school."

She goes back to Nicholas and urges him to send her an e-mail about his new girlfriend, to fill her in on the details of his life so she can participate in some way. But he keeps the same semi-noncommittal stance, and it's not clear if he's just awkward on the Internet or back to his old resentful feelings. And just as Evelyn types out a clipped good-bye, bursting with contained frustration, her ICQ program chirps again. It's Dave again. "Hi handsome, perfect timing."

Evelyn turns the laptop so I can see the letters as they appear on the screen: "Hi beautiful. How are you holding up?"

"Good," she types, filling him in on the surgery news, and finally she gets the response she wants and needs. "Oh that's fantastic my darling," he writes from across the world. And she writes back, "Oh yes, we are on the way home." Then he responds again. "I love you my beautiful woman you are everything to me. I just want to hold you in my arms right now and hold you real close and stroke your hair." And she smiles and writes that her heart is singing because her wonderful man is here with her, keeping her safe and supporting her. "I love you Dave," she tells him. "I love you too my wonderful brave strong woman," he answers. And they continue to type variations on this theme for the next hour, until Dave has to go to work.

. . .

Since I'm in Baltimore, I shoot over to Johns Hopkins to meet Dr. Ain and get his version of the decompression surgery last year. It has taken this long and Evelyn and Jocelyn's written permission to get him to agree to an interview, and he still seems oddly grumpy about the whole thing. He digs out her X-rays and sticks them in the light box on his wall, turning it on with an impatient snap. "Those are her screws."

There are twenty-two of them, long and black in the shadowy gray bone.

"Wow. And what's this rod they're going through?"

"That's a titanium rod that connects the screws."

The whole huge contraption goes up her spine like some futuristic armature, a living metal cèntipede as drawn by H. R. Giger. Ain tells me that his phase of the operation lasted from 4:00 in the afternoon till 2:00 in the morning, ten hours of mortally dangerous surgery under X-ray guides while sweating in a heavy lead gown. He gives the details with professional detachment. But when I ask him if Jocelyn was really (as Evelyn always said) one of the worst cases ever, his lip curls and he shakes his head dismissively. "I've seen a lot worse," he says dryly. Then he asks me to turn off my tape recorder. Jocelyn hasn't been to see him since the operation, he says. He's had no opportunity to do follow-up of any kind. He called four times to ask about new X-rays, only to have Evelyn tell him that Dr. Kopits had taken some and they were fine.

"Is this because of your fight?"

"She told you about that?"

"At length."

"What did she tell you?"

"What I recall is that she felt that you were, you know, a typical arrogant doctor. And she was concerned about whether you could be physically up to it."

Sitting at his desk, Ain looks powerful enough, with broad shoulders and a muscular chest.

"I might be wrong about the arrogant doctor thing," I say.

"No. No. You were right." Then he considers for a moment and decides that since I do have Evelyn and Jocelyn's written permission, he will tell me *exactly* what happened. When Evelyn saw Dr. Carson last July on that frantic trip up from the LPA convention, Carson told her she needed the operation but didn't really specify a time. Later he brought Ain on for the first phase, and Ain passed on the message that he'd need a special series of "plane" X-rays. Next thing he knew it was September, and Evelyn and David showed up in his office with the idea that surgery was a week away. "First I ever heard of it," Ain said. They got angry with him, accusing him of misleading them, and Evelyn insisted that the surgery *had* to be done right away. Maybe they'd actually gotten misinformation from someone, maybe she'd just taken a scrap of information and picked a date and tried not to think about it after that. Either way, the soonest date he and Carson could coordinate was three weeks away. And everything seemed fine until a friend showed him a note Evelyn wrote about him on some kind of web page. "Basically, she lambasted me. She said, 'Dr. Ain was arrogant, Dr. Ain was unsure of himself, I don't know if I should trust him.' And she said, 'Dr. Kopits said the surgery that Dr. Ain's planning to do is going to paralyze Jocelyn.' So at that point I called her up and said, 'Hey, I don't want to do your surgery if this is how you feel about me.' "

That's when she brought up his stamina. "This is what kills me—Evelyn, whose daughter is an achondroplast, had more concern about me than anybody else." Which seemed particularly unfair because his parents never took his being short as an excuse, nor did his teachers at Andover and Brown, or the surgeons who trained him, or the average-size woman he married, not even the average-size parents of the many average-size children he's treated. But Evelyn did.

As he continues, I remember David's remark about Evelyn not accepting Jocelyn when she was first born. Since then, Evelyn has told me herself how upset she was, how they left Jocelyn with her mother and went "on caravan," and even went into therapy for a while. And I remember Dr. Kopits's surprise, at that first diagnosis in Atlanta, when she said she didn't know the Australian specialist named Dr. John Rogers. It all seems consistent with this new story of showing up without an appointment, as if she has to squeeze her eyes shut to face the things she has to do. As Ain tells me that he's never even been to an LPA event except as a doctor, that he thinks it's "artificial" to bond over size alone, I can't help thinking of how passionately Evelyn has thrown herself into the Little World. Does she really see dwarfs as broken creatures? Is she still haunted by that line in her nursing textbook about dwarfs making a living in the circus, still driven by the specter of that red nose?

To clear the air, they scheduled a meeting, Carson and Ain and David and Evelyn and another doctor as a neutral face. And this time, Evelyn said it was all just a misunderstanding. Even so, Ain felt uneasy. "If I had to do it today, I would *never* have operated."

Ain continues without prompting, clearly still rattled by the whole experience. "I mean, people question me all the time. I encourage that. I don't want them just to say, 'Of course, just do whatever surgery you want.' That's idiotic. And if she wanted a second opinion, I would have gotten her one. But she said *nothing* to me. If it hadn't been for her writing on the Internet, I never would have known. I've been here three and a half years. Worst case. Without question."

. . .

Out of the blue, after weeks of little contact, Andrea writes an e-mail quoting that fateful passage from my *Esquire* article, the one that began with "they do look wrong, there's no getting around that." And she asks, "Okay, here's my question. If you were asked to write an update to your article, and you came to that paragraph, what would you write now?"

My answer goes on for pages and pages. Yes, those words strike me differently now. The word "revulsion" makes me cringe. I hate the thought of Jocelyn reading that and being hurt by it, and feel a powerful impulse to clarify: I wasn't thinking of your average achon but of some of the very twisted people "jammed into their wheelchairs," as I so sensitively put it. But I was describing a moment in time and how I felt in that moment. The feeling was strongest at first glance and wore off. I felt ashamed of it pretty quickly. That said, I stand by what I wrote. And maybe I should have put in a "sometimes" or "occasionally" or "initially," but that would just be hedging to protect myself when the fact is, if you glance across the statistical room and see fifteen thousand people with "average" proportions and ONE who's a dwarf, which one looks wrong? Can I be Mr. Rogers? Which of these things does not belong? Again, I'm not saying IS wrong. I'm saying LOOKS wrong. At first. In the reptilian brain. My reptilian brain. Which, as previously mentioned, brings me shame. And maybe Dr. Kopits or Mother Teresa or your good and noble friends don't feel this way or maybe they just don't remember how they felt at first glance, but didn't every damn dwarf at that convention say they felt the same way the first time they walked in—those people look wrong, do I look like that? And there are these scientific studies about "the shock of disability," so I *know* it's not just me. And the funny thing is, I have always preferred the abnormal, at least in music and literature. Love surrealism, expressionism, magic realism, "degenerate art" of all kinds. Which is what I meant by the "liberating delight" line—the idea that exploding normative thinking can set a person free of interpretive screens and mental constructs and the division between self and not-self and Buddhist stuff like that. And maybe you don't like to see your divine and independent self as a mere vehicle for someone else's enlightenment, but I still feel that the bottom line here is this: You are punishing me for telling the truth about things you don't want to think are true.

. . .

One night a strange thing happens. I'm flipping through my journal and the word "dwarf" jumps out at me. I stop and read:

> *Dreamed that I went over to A.'s house and she was going to sleep with me but she wanted me to meet her dwarf son first—then I didn't want to sleep with her and take on the obligation, but also didn't want to be the asshole. I was stuck.*

It was written one year before the LPA convention.

This sends me back through my old papers and notebooks. I find a story I wrote my freshman year in college about a Vietnam veteran who romances a hideous girl out of some dark obsession with injustice. There's another story about a man who is still a virgin at thirty-five because he feels too repulsive to love. There's even a short story about a group of oddball actors staging a theatrical version of Todd Browning's *Freaks*. Then I dig out the magazine article I wrote about Susan Cabot, the actress who got killed by her dwarf son, and find myself startled by what a fanatic mother she was and how desperately she tried to nurture her "poor feeder," how she hid him away from the world in a house overgrown with brambles. Just looking at this stuff embarrasses me, and not because it's so morbid and romantic. Where's the connection? I'm tall and healthy and my own mother was anything but fanatic. Why am I so interested in this? Various theories come to mind, ranging from general teenage alienation to the influence of writers like Fyodor Dostoyevsky and Samuel Beckett. Then one day I'm interviewing a dwarf porn star named Bridget "The Midget" Powerz about her unusual career choice—she started going wild in high school, when she found a home with the punks and rockers and decided there was no point in even trying to be normal—when she mentions getting stuffed in trash cans in middle school. Suddenly I remember with a jolt that there was a dwarf kid in my middle school too—his name was Jerry Goldsmith and he was in my circle of friends, although not a particular pal of mine. And one day someone stuffed him in a trash can. I can picture him upended in the metal can in the locker room, see the moist green tiles and the heavy school ring of the teacher who may or may not have been the sadist who put him there. I don't think I actually saw it happen, but there's this image in my brain. And it's certainly odd that I'm just remembering it now. But I'm pretty sure that this is not the story behind the story behind the story, that the answer can't be anything so simple or specific. I don't know what the answer is.

. . .

As the month goes on, Evelyn sounds at once sadder and more numb with each passing day. Jocelyn can barely talk, her throat is so ravaged by the oxygen tube the nurses push down to her lungs during surgery. They fall into a pace of one long horrible day in the operating room and a couple of nights of almost no sleep and then a few days of recovery and then another long horrible operation day. For three days after each operation, Jocelyn can't eat hard food. By the third round, she hasn't eaten for a total of nine days and her body is going into a negative nitrogen balance. Which means she's beginning to consume her own flesh, and it turns out she's going to need the hip operations, too, despite what Dr. Kopits thought at first. Then she goes in for round four and when they open up the cast, they find a skin infection and have to postpone for a week. The setback hits Evelyn hard, and she fires off a letter to Alecia letting out some of the feelings she has been bottling up. They don't write, they don't call, they don't seem to care at all about the terrible things Jocelyn is going through. They just go on with their happy little lives. The next day Alecia writes back an anguished note:

> *This makes me wanna cry. . . . Mum how can U think that dad Nic or I don't think about Joc . . . WE ALL MISS HER sooooo much. BUT WHAT IS THE point in writing every single day . . . what are we to write?? Everyday the same msg . . . goodluck! U HAVE hours to sit on the computer and write e-mails. We don't have so much time!*

Meanwhile, Evelyn keeps getting lengthy e-mails from her Internet friends who share the most uplifting feelings and tell her she's so brave and inspiring. For example, this note from a man she knows only as Jet:

> *Dear Evelyn and Teddybear, I read your newsletter #3 and i had so many mixed emotions. As an adult with good health, I do not realize how fortunate I am to have the health I do. Sometimes I get mad at myself for being depressed. I have so much to be thankful for. So I guess I am trying to say I admire you. I too went through a divorce and separation, and this was the most difficult thing in my life. For a while, my daughter hated me and yet I could not understand why because I had always lived for and did everything for her. So if anyone has been not supportive of you and even critical of you for doing what had to be done, then they have never really experienced what you have. You have a young daughter that is so full of desire to enjoy a fuller life. With her determination and a mother's love that nothing can match, I just know this will happen. I envy you both and pray for you.*

With this, Evelyn includes a note, marveling once again that "someone you have never met, never seen, never heard, can provide some honesty, some perspective, some support, some understanding." But reading the two letters side by side makes me angry with her all over again. *Obviously* it's easy to be decent and wise with strangers. *Obviously* the hard thing is to be honest and understanding with the people you actually love. Can't she hear Alecia's anguish? Can't she feel her pain? Is this terrible "focus" turning Evelyn into one of those ruthless mothers in old Greek plays? Is this the real reason for our fear, the thing we suspected all along, the secret cost of difference?

Then I remember again how much she is suffering. And how much more misery waits ahead.

Thirteen

THE HIPS TOOK TWELVE HOURS AND THE RIGHT LEG TOOK ten, and this—Jocelyn's fourth operation in little more than a month—will nudge her past thirty hours under the knife. Standing beside the gurney as we head down to the operating room, Evelyn brushes Jocelyn's hair back off her forehead. It seems so pale and vulnerable, like a small pregnant belly suddenly exposed. I never noticed the dramatic widow's peak before. Then the nurse parks us in the hallway and we wait and Jocelyn tells me that her dad called the other day and discussed the weather. Her tone is dry and cynical. After an hour's wait, they give me a pair of green pajamas and lead me to the locker room to change. I'm nervous because it took two visits and weeks of letters and calls to get them to let me observe the surgery, and I made all kinds of promises about how cool and professional I am. But I've never done anything like this before.

By the time I get to the operating room, the anesthesiologist (who is more tolerant of reporters than the last one) is giving Jocelyn her first dose of Diprivan propofol. Once she's asleep, they swing her from the gurney to the surgical table and stretch her arms and legs onto black vinyl padded arms that lock onto the operating table like some high-tech crucifix. Then they pad some more to adjust for the natural inward curve of her arms and wrap both arms in corrugated blue foam that looks like soundproofing material. The anesthesiologist lifts her head and slips a plastic jelly "donut" underneath and tapes her eyes shut to keep them from being scratched, then slides a breathing tube down her throat. A plastic bellows begins to inflate and deflate. Then it's time to hook up the machines that monitor her heart rate (pads glued to her chest), temperature (a tube down her nose), blood pressure (a custom-made cuff for short arms) and the level of oxygen saturation in her blood. There's even a machine that tracks the "minimum alveolar concentration" in her isofluorine vaporizer, whatever that means. All of this takes a full hour. Finally they cover Jocelyn's face with a blue sheet and clip a second sheet to stainless steel poles, creating a blue wall that divides Jocelyn at the waist. Her left leg is the only part of her that remains exposed.

Then Dr. Kopits begins examining the leg. He probes it with his thumbs and strokes it and stands back and studies at it from different angles, then crosses to the other side of the surgical table and studies it some more. For fifteen minutes he does nothing but stare at Jocelyn's leg. He's not just looking at the bone but at the future of the bone, trying to figure out how to compensate for future growth without creating a "parallel deformity." This requires special knowledge of dwarf growth patterns and forces Kopits to use invented procedures or apply standard procedures in unusual ways. Then he puts on his lead apron and cues the X-ray technician, who slides a huge white tuning fork around Jocelyn and projects ghostly images of her hip and leg into a video screen next to the surgical table. Kopits stares at that for a while, then looks back at Jocelyn's leg and back at the screen. To my eye, Jocelyn's thigh seems straight enough. Below the knee is where the dramatic curve begins, sending the foot askew at what looks like a thirty-degree angle. But Kopits sees things much differently. "Don't look at it as three-dimensional," he instructs. "Look at the planes. If you look down like this, the leg is straight. Now, come behind me, and the knee goes straight to the ceiling. Now, if you can lower yourself to put your head right here, I'll show you what's happening—you see the torsion?"

What he means is that Jocelyn's ankle twists like a cheese stick. It also points sideways—and upwards too.

At this point, the scrub nurse begins washing Jocelyn down with a sponge soaked in an iodine wash called Betadine. Carefully, she works the sponge between each toe. Watching her work, so focused that she seems almost reverent, I'm struck by the absolute integrity of skin. It's so sheer, no cracks or grain or bubbles, such a soft and perfect armor. It seems almost holy.

Then Dr. Kopits slips on his fiber-optic headlight, which looks like a high-tech miner's lamp, and goes into the scrub room to wash his hands. While he's gone the nurses "define the edge of the sterile field" by stapling the blue sheets to Jocelyn's leg—actually staple them right to her skin with what looks like an ordinary desk stapler. Soft classical guitar comes through overhead speakers as Kopits returns, holding out his hands like the Frankenstein monster while the nurses pull on his surgical gloves, which are thick and dark green rather than the thin flesh-colored ones you always see on TV. Then he wraps gauze around Jocelyn's thigh and tightens an Ace bandage tourniquet around the gauze, fortifying the sterile field. Carefully, he folds the gauze back over the edge of the tourniquet.

And finally they are ready to begin. Kopits sits on a low stool, eye-level with the left side of Jocelyn's left leg, and the nurse hands him a purple Magic Marker. He draws a line down the outside of Jocelyn's calf. The nurse hands him a scalpel. With one sure move he draws the blade down the pale flesh. It parts so easily, without the slightest tug or catch, spread by the pressure of the yellowish fat below. There's no blood. Kopits moves the scalpel again and this time it goes deep and a few tight red beads ooze into the bottom of the cut. Down there I see a glint of silvery white, which I assume is the bone. The nurse hands Kopits a long electric needle and he touches it quickly to the areas where blood is seeping. There's a slight smell of burning flesh and a wisp of smoke and then the blood stops. Kopits wipes the needle against his wrist-pad to clean it and the needle sizzles. The nurse slips a suction tube into the cut and vacuums out the line of blood, then Kopits takes up two flat pieces of steel, each about an inch wide, curved at the ends, and hooks them into either side of the opening. He pulls and the wound opens. One inch, then two. He runs the electric needle along the bone. It hisses. He works a finger into the opening, follows the fingers with forceps, separating flesh from bone. There is nothing delicate about this. Finger and forceps work rough and fast. He asks for the scalpel and extends the incision two more inches and then goes back to work with the forceps. The oddly bloodless wound gapes open, held back by the stainless spreaders, the geological gradations of skin and fat are all visible. The long line of exposed bone is surprisingly thin. A toddler could snap a twig that size. Another application of the spreaders and a few snips here and there and then Kopits starts to cut at a silvery length of tough ligament-like material, exposing a stem of red muscle that looks exactly like flank steak. As the nurse holds back the muscle with an instrument that looks like a curved fork, he snips carefully at the silvery sheath. I was wrong, it's not the bone. It is called the "fascia" and it keeps the muscles in place. When it's young and taut, it gives our bodies that healthy, toned look. But in an operation the traumatized muscles swell up and press against it painfully, so Kopits is doing a "preventative fasciotomy to give them swelling room." When he's finished, he points into the cut—he's exposed a half inch of skin and fat and two layers of meaty muscle, and down there at the bottom lies the bone. But it is still covered with a thin shiny film called periostium, so Kopits scrapes at it with a scalpel and a small set of tongs, working vigorously and pausing every now and then to pick off the little periostium shreds that cling to the bone exactly like those little white shreds that cling to a T-bone. Which is, after all, what they are. And when he's done with that, Jocelyn's bone is as white as white can be. This is it, the source of all her pain.

Suddenly there's a gush of blood. Quickly, Kopits stuffs a bandage in the corner of the wound. As the orthopedic technician holds the wound open with two large spreaders, he places his saw—a 3M minidriver, which looks like a silver gun tipped with a small blade—against the bone. The saw whines and raises a spray of white dust. It takes only a moment to sever the bone, which gives slightly as the saw comes out the other side. Kopits moves the blade down an inch to make another cut, then snips away a few more shreds of periostium and probes the wound with tongs. He extracts a chip of bone and drops it into a small stainless dish.

Then Kopits stuffs a fresh bandage into the gap and pauses to study his work. After an intense minute of absolute concentration, he calls "scalpel" and quickly extends the incision down to Jocelyn's ankle, stuffing in another bandage (leaving a corner of cloth hanging out as a reminder) and holding the two edges of the wound closed while the orthopedic tech puts in a few staples. This will keep it from bleeding too much while they tend to other jobs. The anesthesiologist notes Jocelyn's heart rate and blood pressure, urine output and oxygen saturation. At a spike in blood pressure, he gives her a little bit more of a synthetic narcotic called Fentanyl.

Forty minutes have passed since they put on the lead aprons. Without pause, Kopits moves to the other side of the leg. With the purple marking pen he draws another line along Jocelyn's skin and follows it with the scalpel. He doesn't hesitate and doesn't even seem particularly careful, jamming his finger into the cut and working things clear. This time he dabs the bone with a line of blue paint. "This is very exciting," he tells me. "See how the grain of the bone twists?"

The blue line makes it vivid, plumb to a bone that swirls up and around it like the horns of one of those wild mountain sheep, twisting the ankle and foot off in an odd angle. Kopits starts the minidriver humming again and when he's through, Jocelyn's ankle rolls loose and free. Then he positions a stainless steel plate—basically just a strip of polished metal with eight predrilled screw holes—and clamps it to the bone, then picks up a small tapper drill and bends close to drill the first hole. The drill comes out a little bloody, a little bit of bone meat stuck in the threads. Kopits hands it to the nurse and begins to twist in the first screw. The whole process is surprisingly like carpentry, and rough carpentry at that. Except bone is alive. It may seem hard as wood but put a screw in and tiny cells called osteoplasts start to reabsorb the part of the bone that's under pressure. The bone melts and the screws go loose, which is why you use a cast during recovery. Then other cells called osteoblasts come along and firm things up again.

The second screw brings the leg into position, sending the blue lines off in different directions. Kopits asks for an X-ray. He waits while the X-ray tech gets the machine into place and the picture up on the screen, then studies the picture for a moment. Turning back to Jocelyn, he takes her leg in his hands and gives it a dramatic twist. And suddenly her foot is straight. I expected it to take more time and healing and growth, but here it is in an instant, like magic. But Kopits shakes his head, once again studying the image on the X-ray screen. "It improves a little bit, but not completely," he says. "Hold it there. Give me the drill." Working quickly, he adjusts the clamps again and drills a fresh hole near Jocelyn's ankle. Then he drills another hole and another, putting in four screws altogether. On the X-ray screen the screws appear as solid black masses in the ghostly bone. Kopits asks for "before" shots and the X-ray tech punches a few buttons, summoning up the old twisted bone.

"Much better, much better," Kopits says, studying the screen. He holds Jocelyn's foot in his hand and pivots it around, studying how it moves. "The tibia's too long, still."

At this point, they have to release the tourniquet for a while to let Jocelyn's blood circulate. During the break, Kopits plans another cut in the bone, lower on the ankle. That will make three osteotomies in one day, he says. "It's like having a major car accident in three installments." And suddenly a memory comes rushing back—when I was seventeen, a car blindsided my motorcycle and sent me to the hospital. I woke up two weeks later with my hip smashed and left leg broken in three places. A triple compound fracture. Three osteotomies. They put a steel rod through the bone just below my knee to keep me from moving, and another steel rod down the center of the leg, and I stayed flat on my back for the next two and a half months. I remember the cats foraging in the Dumpsters below my window. I remember a nurse coming in one day with blood on her white uniform. I was exactly the same age as Jocelyn is right now and it was exactly the same operation Jocelyn is having, with a few more random bone shards. Odd that I haven't thought about it before.

When I turn the tape recorder back on, I am asking Kopits how he can cut so fast. "I know *exactly* what I want," he says. "I have it in my hands. And that machine"—he points at the X-ray machine—"doesn't tell you the truth. Because it cannot distinguish rotation. Torsion it cannot distinguish. It projects everything on a flat surface. It gives you a shadow of something. That's what an X-ray is, a shadow. So it lies."

"You've been down fifteen minutes," says the nurse.

Without reattaching the tourniquet, Kopits picks up his drill and puts in another screw. Blood seeps onto Jocelyn's open flesh, bright against her iodine-yellow skin. Kopits drills quickly, patting the bit dry on a gauze pad. When he's finished, he takes a close look. "The rotation has been corrected," he says.

Three hours have passed since they began. Now Kopits peels back the blue sheet to expose Jocelyn's foot and suddenly the "surgical area" isn't just a slab of anonymous flesh. I can see where Jocelyn's shinbone widens to her foot. I can see her toes. And Kopits makes a fresh cut and exposes two inches of bone and marks it with the blue paint and leans in with the minidriver saw, and this time it's hard to watch—there's more *information* to a foot. It's not just a slab of meat. It that why deformity strikes us so hard? Why it's such a violation? Because it messes up the sacred codes that make us human, that makes our slabs of meat somehow different from all the other slabs of meat?

This operation has been going on too long.

This time as Kopits scrapes the bone clean, he leaves a shred of meat hanging down into the cut. He seems to have forgotten it, or maybe it's not pertinent.

It bothers me.

This time he uses an ordinary saber saw, or something that looks very much like one. As the saw howls and Kopits drives it almost through the ankle bone, stopping just a thumbnail short, Jocelyn's foot relaxes and a triangle of empty space opens up in the cut part of the bone. Kopits squirts a jellylike liquid into the hole, then picks up a bone fragment and uses his fingernail to scrape a little meat off it before slipping it into the triangular gap. "This usually causes a problem at Resurrection," he jokes.

Kopits's carpentry seemed rough before, but this carpentry would be downright bad. Quickly, he drills a few more holes and hammers in a big thick staple, then steps back for a look. On the X-ray machine, we can all see the chip of bone just stuck there, with gaps big enough to run a pencil through. But Kopits is so pleased that he asks the nurses to get me a lead gown and surgical gloves so I can lean in for a better look. Putting one hand on Jocelyn's leg for balance, I bring my face eighteen inches from the—let's call it an "incision" just this once—and see how much more vivid the colors get, the pink and red and yellow flesh and the white white bone.

Kopits points out the triangle. "You see, it's an incomplete osteotomy. It's not quite broken through."

"Yes, I see."

"It creates a perfect tension between the staples and the bone. Now the knee is perfectly lined up with the maxilla, the top of her foot."

They pause to sing "Happy Birthday" to the anesthesiologist, and finally Kopits begins to close up the wounds, first stitching a row deep down in the cut and then a second row higher up. Blood puddles at Jocelyn's heel, soaking the padded sheets. For the final row of stitches, Kopits uses a smaller needle and a pale thread. The skin comes together so clean and tight it is almost seamless, the gaping wound now a fine red line. This is called tri-level stitching, a plastic surgery technique. When he moves to the other side, that gaping wound is almost shocking in contrast. Kopits marks the skin with blue paint to keep the lines straight and drops little chunks of bone into the gap left between the two ends of cut bone, then he takes a drainage tube with a spike on it (exactly like the spike fishermen use to thread a fish line through a fish's gills) and pushes it through the flesh just above her knee. With one stitch, he fastens it to her skin. Then he begins sewing.

"Gonna do the cast too?" the anesthesiologist asks.

"Yeah. Tie-dye."

Six hours have passed since they started. The nurse wheels the surgical instruments away and comes back with a pile of plaster strips, basically crumbs of plaster stuck to rough muslin. Dipping a strip into the water, she hands it to Kopits, who places Jocelyn's foot against his chest and smooths the plaster on with loving strokes. The subject turns to names. "There's a Dr. Foot, isn't there?" asks Kopits.

"Sure there's a Dr. Foot," says the nurse. "There's a Dr. Balm. There's a Dr. Pain. There's a Dr. Slaughter."

And finally the blue sheets come off. I watch Jocelyn's rib cage rise and fall.

It's the most normal part of her, unaffected by surgery or dwarfism. The distortion begins at her hips, driving her thighs into saddlebags so exaggerated they would have driven any woman I ever dated crazy. But her ribs and her belly are perfect and right now, after the violations of surgery, they seem very beautiful.

"She had a fifty-degree flexion deformity," Kopits says.

"She was sitting when she was standing, essentially," the anesthesiologist says.

Jocelyn's chin puckers up. She's coming around. Quickly they roll her onto her side and Kopits cuts the bottom out of the cast, folding back the underlying cotton and coating it with a light layer of plaster to form a smooth ridge. Then Jocelyn starts vomiting and the anesthesiologist sucks her throat clean with a green suction tube.

"Do you want her on her back?" Kopits asks.

"No, this is actually the absolute perfect position."

A few minutes later, the nausea passed, Jocelyn opens her eyes a slit. She looks exhausted. As she lies there, limp and barely conscious, the nurses start to drip orange and blue dye onto her cast. "These are trained professionals," someone jokes. "Don't attempt to do this at home." Kopits lifts a little bottle of red dye up high and squeezes. "See the splash?" Kopits says. "The splash is the thing. It takes finesse." While they drip and kid each other, enjoying the relief at the end of the long operation, Jocelyn's lip trembles. Then her chin trembles, and a tear leaks down her eye. The shiver goes down into her body and she stares up at the ceiling. Noticing my look of concern, the anesthesiologist tells me she's not really cold. "She's thirty-seven degrees. It's probably hormonal." He gives her an anti-nausea drug called Zofran.

I notice that her left arm is hanging out, her hand dangling, and I feel bad for her lying there so sick and traumatized, so I take her hand and give it a squeeze. A moment later, with a one-two-three and a gentle heave, we move her back to the rolling gurney and Kopits bends over to give her a smile. "Beautiful result, my dear," he says. "You have gorgeous legs."

. . .

In the recovery room, I bend over the gurney. Jocelyn's bangs look red against her milk blue face. "You look good," I say.

"I just want the pain to go away."

Hearing that, Evelyn interrupts her post-op with Kopits. The nurse is more important right now. "How often can you give her Fentanyl? Every fifteen minutes?"

Kopits lingers, chatting with the head nurse. When she takes a call, he reviews the operation for me, describing the special problems of these prolonged multiphase procedures. "After the third surgery, you begin to digest your own body," he says. "Sometimes the bones get so weak, you start drilling and they just fall to pieces. There are also emotional effects like depression." Since he doesn't seem in any hurry, I ask him about something that's been puzzling me. As he operated, he pointed to Jocelyn's leg and said, "Here is the deformity" and worked so hard to correct it. Instead of turning the stitches over to an assistant, he spent more than an hour doing meticulously trilevel stitches so his carving wouldn't leave an ugly line. And he was so happy telling Jocelyn she had gorgeous legs. But what about beauty? What about the aesthetic deformities? Why does he have such contempt for leg-lengthening, for example?

"You shouldn't burden an individual for being short by subjecting him to multiple surgeries, okay? Punishing him because he's short—that, to me, is inexcusable. And in some countries, that is the only treatment that little people get."

He stops and starts over, softer this time. Really, the problem is ignorance, he says. The fear of difference is caused by a kind of cultural isolation. "Let's say you and I would be shot up to a planet of little people, okay? And we come out of our capsule and they say, 'They're much too tall for us. Let's shorten them a bit.' " When I bring up the plastic-surgery stitches again, he denies doing them for cosmetic reasons, insisting that it's mostly for function. Once you start something, you might as well do it perfectly. "You see, I wasn't born in this country," he says. "But I'm very much an American for what that means concerning acceptance of differences. This is really the most tolerant country in the world in terms of differences—ethnic, religious, political, etc. This country was built on that. And I'm all for it. To have your son shunned from acceptance to university because he's a little person is horrendous. But the blame is not on your son, but on society. Let's change society."

Admirable as these sentiments are, there's something missing. "Maybe this is off the point for you," I say, "but it seems to me that a lot of their suffering, in this world, is because of cosmetic reasons."

"I fully agree," he says. "I fully agree. Particularly in the face and hands. That accounts for employability." That's why he's recommended surgery for achondroplasts with protruding tongues, because in our society anybody with a tongue sticking out is going to have a hard time getting a job. "But even that has a functional reason. And you can say, 'Well, sure, somebody who has leg-lengthening will be able to reach one more level in the grocery store.' The problem is, at what expense? Okay? If I could, by imposing my hands on the legs of little people, I could make them be whole, be long, I would do it. That's not the question. The question is, at what expense? What suffering will society extract to accept those individuals? And *this is where I rebel.* This is where I say, 'No!' This is where I say, 'It's barbarous.' "

. . .

After Kopits leaves, the nurse asks Jocelyn what number she'd give her pain, ten being unbearable.

Jocelyn clears her throat. "About an eight," she says. "About an eight."

Her eyes stay fixed on the ceiling.

As you sit in Recovery for protracted hours, you see how ordinary it is. It doesn't feel like a crisis the way it does on TV or when you're the patient. Another gurney comes in, they call for morphine, the nurse goes to the medicine safe and punches in the codes. Then another gurney comes down and Evelyn calls David on the nurse's phone, coldly giving him the news. "Oh yeah, she's a tough kid," she says.

Hanging up, she leans toward me. "While you were talking to Kopits, Jocelyn came out of the anesthesia for a minute and the first thing she said was, "Remind me to tell John that I really liked it when he held my hand. I was cold, and it made me feel warm."

That makes me feel really good.

Then we somehow get onto the subject of what kind of car David bought, and Evelyn can't remember the brand name. Then we hear a voice from the gurney. "Hyundai."

She doesn't even open her eyes.

"You're listening to everything we're saying, aren't you?"

"Yes, and I never gas bag. I'll never tell."

She's a little delirious.

A few minutes later, and she opens her eyes and looks up at me. "Thank you for holding my hand," she says. "It really was good." I give her arm a squeeze and her eyes go back up to the ceiling and I get this image of her all dressed up in a dancing gown to go to the LPA ball, a little mascara on those eyes and a splash of lipstick. She's been cute to me for a while now, but today I see the beauty in her high white forehead and fine auburn hair. Or maybe the truth is that I'm starting to *want* her to be beautiful. And I don't care if the mascara and lipstick I'm imagining is just a phony imitation of health, a bit less painful than leg-lengthening but no less slavish to arbitrary social standards, or if flat bellies and firm thighs are just a by-product of the fascia's brief prime. I'm sending her off to the prom and hoping the boys will be nice to her.

By 6:30, her pain is down to five. "We'd be screaming if we were her," says the nurse, "but she just stares at the ceiling."

I ask Jocelyn what she's doing, staring at the ceiling like that.

"Focusing," she says.

. . .

The next morning I arrive in the hospital room to find Jocelyn lying very still, trapped like Lot's wife in that waist-to-toe cast. Her eyes are closed. She has little blue booties on her toes where they stick out of the plaster.

Behind her on the daybed, Evelyn sits against the wall, typing away. "What have you been up to?" I ask.

"Well, we painted a Picasso. We went shopping. We had a walk."

She looks exhausted. She sounds even worse. This is the first full operation for me, but they've been trapped here in this room for five weeks now. Jocelyn's dresser is covered with flowers and teddy bears and a coffee cup called the "Big Hug Mug." The wall is decorated with cards that say things like, "To help you look on the bright side, here are things that could be worse: Your butt could fall off. You could be a mime. You could like kareoke. You could be one of those perky people who smile all the time." But all this cheer only mocks Evelyn's black mood. She says the news of the day is that David's efforts with the Australian government paid off, that they're going to pay the rest of the hospital bill. "Don't congratulate me," she says. "I had nothing to do with it. I was a big stumbling block."

"Maybe we should go out for lunch. Give you a break."

"I don't want to leave her," she says.

"Evelyn, we're in a hospital."

"I know," she says. "Surrounded by trained professionals. Don't give me a hard time. I don't need it."

She goes back to her laptop.

. . .

Jocelyn's roommate is Dawn Lang, the little Pamela Anderson look-alike I saw at the track my first day at the LPA convention. She's also seventeen, but not the same way Jocelyn is seventeen. Her hair is carefully streaked and her rolling lap table is covered with makeup and hairpins and a hairbrush and Pantene Hair Conditioner and a copy of *In Style* magazine. Her mom had to go back to work, so she's staying here alone, no big deal. She's had three operations "and a couple more to remove hardware" and she's home-schooling because of the surgery but she's way ahead in most of her subjects and wants to be an oncologist or a surgeon or possibly a nail technician. Her fingernails are painted green.

"Jocelyn's really nice," she says. "I wish she talked more. She doesn't really say much."

When Jocelyn wakes up, Dawn asks her if she dyes her hair. Jocelyn says she does and Dawn smiles happily. "What do you do with the roots?" When the conversation lags, Dawn notices Jocelyn sipping at her hospital juice and that's enough to keep her chirping along. "Do you like the orange juice here? I think it's so sour."

Jocelyn barely answers. Looking at them there on the beds, both with short arms and short legs and big heads, you can't help wondering what made them so different. America? The wheelchair? Maternal styles? A few critical degrees of beauty?

Evelyn sighs. "Isn't she a breath of fresh air? She's so perky."

. . .

Sitting across from Evelyn, laptop to laptop, I write Andrea a note about Jocelyn's surgery and the flashback to my own surgery and how peculiar it was that I could feel for Jocelyn a sorrow that I never felt for myself. Andrea writes back: "It's great that Jocelyn's experience reminds you of your own. But stop pitying her. I'm not saying, don't feel bad for the pain she feels—and while you're at it, for the pain you felt—but don't fucking idolize it."

. . .

Evelyn has discovered a new Internet gizmo called Mplayer so she and Dave can put on headsets and hook up in a kind of chatroom and talk live. But it's slow and frequently interrupted, so she asks me to buy her some long-distance phone cards, the kind that give ninety-seven international minutes for twenty dollars. When I ask her how many she uses, she tells me she runs through one or two a day. Mplayer doesn't offer enough privacy. I do a little mental math and open my mouth to comment.

"Don't say anything," she orders. "I have to talk."

I shut my mouth.

"Good," she says. "Whatever it was, I don't need to hear it."

. . .

Just before dinner, Dr. Kopits comes in wearing a Winnie the Pooh tie. He stops at Dawn's bed. "How's your pretty foot? How are you doing?"

"Is that normal, to turn purple like that?"

"Is it normal to have green nails?"

He comes over to Jocelyn. Evelyn asks if he's going to have to use anesthetic to take the pins out of Jocelyn's ankle, and Kopits tells her that Dawn did it without any drugs at all.

"Just so long as you have all the novocaine and everything, it's not so bad," Dawn says.

. . .

Today Jocelyn's feeling better. Although the cast pins her flat on the bed, she's watching TV through a pair of prism glasses, thick mad-scientist wedges that tilt the image of the screen down to her upturned eyes. When I tell her I want to take her shopping when she's ready to buy a gown for her first LPA ball, she lets slip the first complete sentence I've heard from her since the operation:

"You'll be shopping for a long time, 'cause I'm very picky."

In these days of recovery, I feel closer to Jocelyn than ever before, partly because of the way she thanked me for taking her hand in the operating room. Until that moment, she seemed so remote, and it's taken me all this time to realize that she's been focusing her whole life just as she focused on the recovery room ceiling. Then somehow, when I was there *with* her, when I took her hand and later when she told me it was good, we walked together across the bridge of pain. Or so it seemed to me. And now it strikes me that pain is the key to all of this. No wonder David is the odd man out. No wonder all the trust goes to Evelyn. Now that Jocelyn and I have been through surgery together, I'm getting the tender smiles and the trust. And it feels good. It feels really good. And I can't explain all the reasons why it feels that way, but I know that I want to take it deeper. So I tell her the odd thing I remembered in surgery, that I had the same triple fracture at exactly the same age.

"But did you do it on purpose?" she asks.

. . .

Every night, back at the Pierre House, I run into a mother smoking cigarettes on the porch. Her name is Kiki and she's here from Florida with her eight-year-old daughter, Michelle. Michelle has SED. Not only was she born short, she was also born without heels. Without surgery she would never be able to walk normally. At first Kiki thought God was punishing her. She was just getting over that when Michelle's dad distanced himself right out of the picture. "He couldn't handle it," she says, with matter-of-fact bitterness. After that, Kiki went through two years of silence and withdrawal. But she got over it. And she's learned to be grateful. Because the surgeries have been hard but that little girl in bed next to Michelle is three years old and wore a halo brace drilled right into the bone of her skull until yesterday, and when they drew the curtain between the beds and took it off, Kiki heard through the curtain the father talking about how long it had been since he'd been able to hug her. So you better look around when you feel sorry for yourself, because there's a long line behind you. And anyway the important thing is not what you do for dwarfs but what dwarfs do for you. Like the way Michelle never let her disability stop her, the way she learned to run around everywhere on the balls of her feet, dancing and flying around like she was too light to settle on solid ground. "These kids are amazing, they really are. And *they* change *you.* The way you see yourself, and life, and everything. You start falling in love with your child, and one day you just get up and you're a nicer person. You're more compassionate. And you're happier too, just from seeing her struggle, getting up every morning, going to school—a straight-A student."

She starts to cry.

. . .

"Where are you on the pain scale?"

"Four."

"How are you doing?"

"Okay. Day Two."

"So, are you bored?"

"No."

"No?"

"No."

Whenever Jocelyn needs something, when she can't feel her toes and wants someone to touch them, when she wants fresh water, when the sheet bunches up under her and needs to be pulled smooth, her eyes go to her mother.

We try to remember the names of the seven dwarfs. We can remember all except one.

"Whatever it is," she says, "I'm Dopey right now."

Down on the daybed, Evelyn puts on her computer headset and makes contact with the Davenet. "Hey, handsome. You have to keep your finger on the button for just a little bit more. Yes, I hear you. I hear you too. I love you too. You're pretty special too. Hello? Can you hear me?"

He's coming to visit in November.

"You are my life," Evelyn says.

. . .

Late last night Jocelyn lost feeling in her feet and they had to call Dr. Kopits and he came with a plaster saw and cut the cast loose and Evelyn would really like to get Alecia in front of her face and tell her a few home truths and maybe she should send her the same angry letter she sent David but then she'd be playing the same game he is, using the kids as pawns, and it's so much worse than last year what with all the emotional issues and unresolved things and she can't wait for Dave to fly over here and now the terrible news is that Kopits is going to have to do surgery on Jocelyn's feet or she'll have to walk with a wedge and that's going to be *another* six to eight hours per foot, two and maybe three operations more and once again they have to find the money to pay for it!! And last night the nurse said they'd scheduled it for December 3 and that's Dave's last night in America and it just doesn't seem fair! "I burst into tears," Evelyn says. "I said, 'Please, please just make it the fourth or the fifth or the sixth, *anything.*' "

. . .

David says that the Appeal Fund fell apart as he expected, destroyed by Evelyn's caprice, but after much persistence on his end, the Australian government did agree to kick in the rest of the cost of the operation. He'll never understand why Evelyn doesn't appreciate that, especially when her daughter's future is at stake, and it's very upsetting always to have to communicate with Jocelyn through Evelyn, who controls the computer, since he's reached the point of avoiding all communication with Evelyn except when necessary:

I am rather sick of her vicious E-mails, full of meta-language and innuendo, and why she sends them I do not know! I have learnt to 'ignore' them and not respond to them at any time except for messages on Jocelyn or about Jocelyn. I am fully aware of Evelyn's circumstances with respect to caring for Jocelyn and constantly remind Alecia and Nicholas of the commitment it takes to do this. Only time will tell for them. As I remind them always, their mother is a talented and wonderful lady and a dedicated mother. BUT I can not and will not defend her behavior towards them. I say nothing. I hope this gives you an insight as to what is happening here.

Right now, David's big hope is that he'll be able to reconnect with Jocelyn in person in February, when he flies over for her eighteenth birthday and her final release from the prisons of wheelchair and plaster.

. . .

From a pay phone in the hospital lobby, I call Andrea. When she picks up the phone she sounds woozy. She says she's got really bad shingles on her butt and she's taking Percodan. "You poor thing," I say.

"You've been so much nicer to me lately," she says.

"That's because I want something from you."

She knows what I mean—the essay she's been writing about me. Last time she mentioned it, it was sixty pages. I've been trying to convince her to let me use some of it in this book.

"I'm not going to send it to you, I decided. Maybe I'll let you read it after you're finished."

"You're such a pain in the ass," I say.

"Do you know you're the only person who calls me that?"

"The whole rest of the world thinks you're incredibly easygoing?"

"Fuck you," she says.

"Send it to me. Send it as an act of friendship."

"Don't try guilt on me. It doesn't work."

"Come on. Be brave! Throw open the doors and windows!"

"I just don't want the exposure," she says.

"I won't use your name."

"I just don't want people to relate to me on the basis of what they've read."

This is her new line. It's not that she's a withholding cowardly tease, it's just that she wants to be perceived as her own pure self, without the hideous screen of my writerly distortions. "Excuse me," I say. "But didn't I just promise not to use your name?"

"You must think I'm stupid. You give it to me in writing that you won't use my name, and then *maybe* I'll let you see it."

"Have you considered that those shingles are poetic justice for being such a huge pain in the ass?"

"I hate you," she says.

. . .

The nurse uses a Hoyer Lifter to get Jocelyn out of bed for the party. "I think we're going to have to turn her—no, we're too high."

Evelyn can't stand it anymore and takes over. "See that knob down there? Turn that. Okay, now, all together."

I take Jocelyn's arm. Because of my clumsiness, one of the chains slips loose and almost hits her in the head. I catch it just in time, with a nervous look at Evelyn. She noticed, of course.

In the playroom, they're having a Halloween party with ghost cookies and popcorn hands with candy corn fingernails. Michelle is wearing a spooky T-shirt and Dawn has her hair done up in cornrows. There's a boy with legs so twisted he has to wear plastic shields around his shins. His hand is shaped like a backward fin. A nurse spots our custom flat wheelchair. "Hi, Miss Jocelyn, finally out of the room!"

As they tell ghost stories and play bingo, Jocelyn lies in her cast and monitors the scene through her prism glasses.

. . .

Jocelyn's watching *Baywatch*, and Evelyn's on the phone. "I love you, handsome. I'll never trade you in. I love you."

When I talk to Jocelyn, I whisper. I have to be quiet whenever Dave's on the line. He doesn't need to be reminded there's another man around, Evelyn says.

An hour later, she's talking to him again. "That's a wonderful idea. I love you, handsome. Oh, you're something. What a wonderful man you are."

The minute she hangs up the phone, it rings again. "I love you too," she says. She puts it back down, sighing. "He's the most romantic man. Every bone in his body is romantic."

Before long she's back on the Davenet, typing away. She's found the perfect escape, leaving only her body behind.

When Evelyn goes to the bathroom, I ask Jocelyn, "How do you feel when she's talking with him and says stuff like, 'You are my life.' "

"I don't even listen," she says.

. . .

Jocelyn is a good patient. That's the phrase Evelyn uses, and it's true. When she's hurting bad, she becomes so soft and unassuming, she's almost saintly. She turns to Evelyn like a flower to the sun.

"Mum, can you take my foot down off the thing?"

"All of it or just the half of it?"

"Half of it. Can you feel my tootsies, too?"

"You're such a worrywart," Evelyn says.

Today her pain level is down to three. The nurse asks if they can take out the intravenous drip.

"Sure," Jocelyn says. "As long as I get the Tylenol Three I'll be okay."

. . .

Sitting on the chair across from Jocelyn's bed, I access my e-mail and find a note from Andrea. "There's nothing fucking inappropriate about our bodies," she writes. "I have a great body. Some people adore my body. You may not like it, and while I'm not thrilled to think that, I'd rather you say that than convert your personal discomforts into statements of fact."

Here we go again. Her latest argument is that by using the phrase "they look wrong" I gave my observation the "status of fact" instead of keeping it safely in the realm of my (hateful, disgusting) personal opinion. And it's not only "achon" bodies she's talking about but all bodies no matter what the deformity. Because the thing I'm just not getting (she says) is that many people don't experience this as a loss. That is my projection of my fear and has nothing to do with them, and she's not going to let me put it on her. "It's like you regretting that you can't move a car, when someone's blocked your car in. It's a fucking frustration, not a life regret." The important thing is that the human condition is similar no matter what you look like, that people large and small have similar angst and similar joys. "And don't read this to mean that I'm saying that difference means nothing. Difference is huge. This feels so fucking frustrating because it seems that you don't listen to me. I know people think we look wrong. I'm the one who gets that reaction with some frequency. But they're not friends of mine. I'm not shaped by their thoughts about me, and if their reaction involves rejection, there's not the loss that I would have if that same reaction came from a friend."

Once again, Andrea insists that the thing I'm not dealing with is what difference triggers for me. "Ten years ago I got into a conversation with a friend (one of you guys) and I talked about my feelings about difference, about feeling profoundly different or feeling profoundly ugly. And she told me that she knew average-size people who felt as different as I did, for a variety of reasons. It was a life-changing conversation for me, because I realized that I wasn't the only one who felt shame, or ugly, or whatever."

But her relief didn't last long, Andrea continues, because she soon realized that this shared sense of difference only intensified the problem. Because when everyone feels different, they treat people who also *look* different as a kind of totem for their own alienation. The sight of someone so visibly unusual triggers their own queasy feelings of freakishness, just as Frankie Addams was transfixed by the sideshow hermaphrodite in *A Member of the Wedding*. And most people can't stand that, so they discharge the disturbance into pity, which wouldn't matter either except that *the person who is pitied begins to feel there is actually something pitiable about him*. And that's what really pisses her off. Because the only thing that's really wrong is all the idiotic assholes—like, say, me—who *think* there's something wrong. "Otherwise, for the most part, it's just fucking frustrating. I mean really, how much of an impact does my shortness, the length of my arms and legs, the size of my head, make in my life, beyond that I can't fucking reach things?"

. . .

Evelyn is organizing. They can't do this new operation on the feet until the ankles have had at least four weeks to heal, so now she has to arrange transport to Atlanta. Which is going to be quite a production, with Jocelyn strapped to a board and laid like luggage across three seats and the Pittsburgh fire brigade on call to help them change planes. Plus she has to outfit the room in Atlanta with a new bed and a lifter and a flat wheelchair. But it's a blessing to have a new goal to focus on. Atlanta! Their little room! Boy, is she dying to get out of this hospital!

She's been acting distant and withdrawn, avoiding any personal questions. Everything she says is on the surface or about Jocelyn. At one point, she insists on putting me on the phone with Dave. He tells me that the thing that caught his eye was her Internet name, Free. He'd just gone through a divorce too, and Evelyn seemed sensible and real. When she told him about Jocelyn, he was intrigued. He did worry whether he'd feel awkward around her, if people would stare or treat him a bit weird, what it would be like with the wheelchair. But that was all over by the time they got to Disneyland. "I love Jocelyn more than anything," he says. "The only person I love more than Jocelyn is Evelyn."

. . .

A real friend would ask a few questions. Like, what kind of person falls in love on the computer? Hops a plane all the way from western Australia? Loves Jocelyn "like a father" after one trip to Disney World? *If this guy Dave is such a loving soul, why is he divorced?* But I don't say any of that. And I don't tell her that what she really needs is a walk outside, to take an hour for lunch, to let the nurses deal with Jocelyn for five minutes and *do something for herself,* because that's exactly what David would say and we all know what happened to him. And anyway who am I to speak? Evelyn is saving her daughter's life. She's a heroine. And if her heroism is not the clean and pretty thing of storybooks but something driven by anguished mixtures of darkness and love, then isn't that just a lesson in what heroism really is? Evelyn is driving herself crazy because that's where she needs to go to get this done.

Fourteen

THE THEORY OF EVOLUTION TAUGHT US THAT A DOG was not a pale copy of the Platonic dog but the sum of all the changes it had gone through, not an essence moving through time but an accumulation of variations produced by time. It took a while for this idea to sink in, but gradually the focus of philosophy began to shift from the essential to the variations—to difference. By the early years of the twentieth century, a French linguist named Ferdinand de Saussure cast a Darwinian eye at language and argued that words themselves had no essential meaning, that language was a system where each piece had meaning only through its difference from all the other pieces. A few years later, a German philosopher named Martin Heidegger began making a series of dramatic proclamations about the implications of difference, which he said were going to blow apart the philosophical enterprise forever. For thousands of years, he said, philosophers had tried to probe the meaning of Being without ever once asking about the difference between Being and beings, between Essence and its infinite variations. In fact, they couldn't even think in the traditional philosophical way without ignoring difference and all its myriad distractions. Which is why they had all gone so terribly wrong. Because difference was the ultimate cosmic skeleton key to everything, "the ground plan in the structure of the essence of metaphysics."

Hard as it may be to imagine, this stuff hit the intellectual world like Elvis in 1956. We had finally arrived where we'd been headed since we left the Garden of Eden, and all of Western civilization had just been one big detour. Through the 1950s and 1960s, the children of Saussure and Heidegger began to romp through anthropology and psychology and even literary criticism, writing books with titles like *Difference and Repetition* and *Writing and Difference* and *Identity and Difference.* Claude Levi-Strauss studied the structures of human societies and decided that human culture radiates in patterns from primal differences like male and female and "the raw and the cooked" (the title of one of his books). Michel Foucault studied the many ways society controls the abnormal, arguing that the more civilized we became the more we needed to exclude certain kinds of threatening human marginalia—that when the leprosy epidemic of the Middle Ages began to fade away, for example, we put the insane into the empty leper asylums because the dawning Age of Reason required an "operation to annihilate nothingness." Then Jacques Derrida extended these arguments to nearly apocalyptic levels, claiming that Western culture itself was organized around the pursuit of Truth in exactly the same way religion organized itself around the presence of God, with the result that the meaningful center is always engaged in police actions against anything that might call its meaning into question—things like death, emptiness, criminality, perversion, wild music and extremely difficult philosophical books. The bitter joke is that the center is always haunted by its margins and the whisper of the excluded fills us with anxiety unto nausea. As a wise man once said, dread is fear of what we desire and desire of what we fear. Derrida's project was to break this pattern first by taking the side of the marginalized and then—so the marginalized don't simply become the new center—by refusing to take any side at all. Throwing all certainty into the air, he welcomed the "pure play of differences."

All this was, of course, infused with the politics of the times, from anti-colonialism to civil rights and feminism and the various sexual freedom movements. It's no coincidence that many of these thinkers were themselves different in some way; Foucault was gay, Derrida a French Jew who grew up in Algeria. In their wake, the old bohemian fascination with subcultures exploded into the world as we now know it, with Broadway musicals about the old circus freak shows *(Side Show)* and movies about Siamese twins *(Great Falls)* and sadomasochism as advertising cliché and academics writing books about the theory of garbage and the history of shit and the "cultural freaks" (as one critic called them) parading their fat and fetishes on the daytime talk shows and kids everywhere using tattoos and piercings and hair dye to make themselves just a little more freakish—and dwarfs seemingly everywhere, from Mini-Me in the *Austin Powers* sequel to Kid Rock's dwarf sidekick to Bridget the Midget to Koko the Killer Klown, who traveled with the Lollapalooza summer rock tour as part of a "postmodern" sideshow featuring a feminist fat lady and a man who dangled heavy objects from his penis. In Hollywood recently, so many people went to an entertainment industry party featuring a performance by a dwarf metal band that the fire department shut it down. In England, a troupe of dwarf male strippers kicked off the new millennium with a successful tour of the countryside, billing themselves "The Half Monty."

And the children of Foucault and Derrida cheer it on. "The return of the freak in millennial culture signals a crisis in the dominant culture's authority to decide who stands in the center and who sits on the fringes," writes a fashionable young critic named Mark Dery. Calling freaks "poster children for an age of extremes," the perfect symbol of the "chaos culture," Dery insists that the whole thing is all in our heads. "Though we stand on the other side of the footlights, we're freaks as well, in the sense that we have internalized the cultural codes that make a freak. In our inability to see extremely tall or short people as just that, our insistence on viewing them in the fun-house–mirror distortions of myths about giants and dwarfs, we participate in a psychology that adequately earns the term 'freakish.' " Only when we realize this, he says, will the "consensual hallucination" that we call normal life finally wither away.

It's the flip side of Naomi Wolf's argument that beauty is just a currency system, a trendy version of the old religious idealism. It would be so pretty to believe.

. . .

These last few weeks, Andrea and I have been getting along exceptionally well, which seems to translate into lots of affectionate ribbing. She teases me about all the delusions I'm experiencing in my "old age," and I tease her about what a pain she is. "You are such a generous man in the face of such childish disrespect," she responds. "I am forever grateful and humbled." Sometimes there are exciting moments when we both seem on the verge of making some kind of breakthrough. One long phone call reminds her of when we met in Atlanta and how much pressure she felt to talk. "At the time, it was overwhelming. But lately, as I'm writing about having a voice and not having voice, I've thought about that a little differently—that in a way, you were pressuring me to have a voice."

But each approach seems to provoke a corresponding recoil, and this exchange ends with another one of her you're-just-using-me-for-your-book notes. "I realized today that to a great extent I'm still a subject for you, not a friend. I'm not important to you, I'm not someone you like a lot/love. I know you probably don't want to call me, but I really need to talk with you. Please call."

Then she asks me for copies of all our e-mails so she can use them in the piece she's writing for her class. This strikes me as more than a little comical, given her resistance to my own writing, so I semi-jokingly propose a trade. I'll send her the e-mails if she sends me the damn story. And if I did use some of it in this book, it would give her a chance to get her point of view across without any interference. I'd even give her some control over the edit.

No deal, she says. "What I've written is more personal than we normally are. An equal trade would be you sharing something with me, about us, that was more disclosing and made you more vulnerable than you have been with me."

I try to rise to the occasion, promising not only to send her the e-mails but to try to be more revealing. I even admit that my habit of keeping a certain distance is some kind of protective device and "probably a failure of some kind on my part." On this basis, we communicate pleasantly for a few months. She agrees to let me use her in this book "minus my name" and even promises to show me her story someday. Then her father has another medical crisis, and she starts sending me long letters laced with anger and raw emotion, all about her father. "The other night we had a conversation about his mind and he said he wouldn't want to go on living if it couldn't be restored," she writes one day. "I asked what about other enjoyment, such as the conversations he and I had been having. He said that it wasn't meaningful pleasure."

Bastard! Monster! No wonder she's so angry! No wonder she's so needy!

That night, I fire off the personal disclosure she's been wanting for so long:

> *No time for a substantive response, but you wanna know what's up with me? I'm in therapy trying to beat back my drug habit, which I now realize is linked to deep anxiety and a sense that I deserved the neglect I got as a kid—as my therapist said, "Funny how you keep using the word 'disgusting.' " This seems to me to relate to what you once said about difference, only difference is too neutral and polite a word. But I really don't have time to go into it right now!!!!*

The next day, she sends a note saying she loves me.

. . .

The night before they leave St. Joseph's, Evelyn looks and looks and can't find the laundry detergent, so she goes and asks the nurse and the nurse comes back to the laundry alcove and points to the bright pink box right there on the shelf.

Three weeks later, she's three weeks worse. They're trapped in their little Atlanta room and Nicholas snapped at her on the phone and Alecia just won't talk on the Internet and they're running out of money and David's not going to contribute—he's sitting there having a good time, hasn't given up his job, hasn't given up a year of his life, and all she can do is fester and fetch things for Jocelyn and dream about Dave coming in two weeks, two weeks, two weeks. "Jocelyn is the best patient in the world, but she has all these *needs*," she tells me. "Sometimes I'm past the point where I remember there is another life outside this room."

It's a survival year, and she's just trying to get through it.

A follow-up note from Andrea:

Hi again. I've been thinking about your e-mail and trying to remember all the things I've said about 'difference.' What this reminds me of is how I used to feel about looking different. Not believing people would actually want to be friends with me, feeling repulsive when I'd look in the mirror. I think the use of the word 'different' is significant, because I believe that all the repulsive feelings I associated with being different were connected but not necessarily inherent to the experience. More simply put, I don't think being different necessarily means being repulsive. I wanted to share that for what it's worth. I'm curious, how do you use the word 'disgusting' in therapy?

. . .

Dave comes and goes, and a week later Evelyn sounds more depressed than I've ever heard her. "I can't talk right now," she says, handing the phone to Jocelyn.

Jocelyn says she's getting better every day.

A few minutes later, Evelyn comes back on. "I had a wonderful time with Dave," she says.

"What did you do?"

"Nothing. There was nothing we *could* do—we had Jocelyn."

"Are you okay?" I ask.

"I'm just down in the dumps."

. . .

Andrea says she's been thinking about cutting herself. It's an image that comes to her when she has strong feelings and doesn't know how to get them out. Like this weekend, when her dad wanted to see her and she didn't go and he said that maybe he should try being more like her, like someone who doesn't care. At times like that, she pictures drawing a blade across her skin and watching the blood seep out. "Don't be alarmed," she quickly adds. "It's only a metaphor."

The next day she sends a note saying she hopes her confession didn't come off weird or scary and that it was really just an image, just a metaphor, nothing more.

. . .

The last time I talked to Evelyn, we agreed that I would come down for Jocelyn's last surgery. But as the date approaches, there's an ominous silence from Baltimore. I send a message and then another and get no response. I can't reach her through the ICQ chat network either.

The surgery date comes and goes. Still no word.

Finally I write a begging note: "Evelyn? What has happened? Please contact me!"

The next day she writes back, a long letter bursting with compressed emotion:

The truth is that right now I am totally fed up with almost everyone in this world, and except for a very small group of friends, I really could care less if I never saw them or spoke to them again. This is not a temper tantrum from a tired little girl. This is fed-up-and-had-enough from a grown woman who has given her all to try and please everyone.

She goes on in fat groaning paragraphs, describing how she's been at David's financial mercy and he's holding up a three-thousand-dollar expense disbursement and hasn't contributed "one penny to Jocelyn's welfare" since they left Australia and she doesn't even hear from him or the other kids "except the very occasional ICQ message inquiring to Jocelyn's condition and maybe the odd phone call and stupid e-mail that he writes to her." She describes the surgery and her fear in the waiting room and what they're going through now—Jocelyn is in terrible pain and can't find a comfortable position, and it's getting more and more difficult to watch her suffer. Every day she wonders where she's going to get the strength to make it to bedtime. "But I

AM going to find that resolve to continue and be damned if anyone thinks I am going to remain being a victim."

And frankly, she says, my questions aren't making things any easier. She feels pressed. She feels exposed. Dave is the only one who understands and all she wants now is to be alone with him:

For years now I have stood up for the rights and freedom of my children, particularly Jocelyn. I now believe it is time for me to do the same thing for myself, to start standing up for myself and demand the respect I believe I have earnt, and to live the life I dreamt of for so long with Dave.

. . .

A few days later, Andrea calls. I tell her I can't talk. I'm about to tell her why, because my mother has just arrived from Mexico and she hates the condo I picked out for her and hates every stick of furniture in every store in New York and hates being a widow most of all. But I can't because she hangs up on me.

She follows it up with a huffy note: "Just want to let you know that I'm probably not going to call Thursday. I'm just not in the mood for chatting."

I write back a note even huffier:

This has been one of the most stressful weeks of my life, I haven't slept five hours a night in a week, my mother had just arrived and was weeping all afternoon and I had two kids and three phone lines working when you called, and you fucking hang up on me because I can't talk RIGHT THAT MINUTE.

A few minutes later I calm down and send an apology, telling her I care for her and don't want to squabble.

Her response arrives the next day: "You're angry at me because I'm not coping? My father is fucking dying before my eyes, and you send me that e-mail now?"

. . .

I write back to Evelyn to apologize for pestering her. "If you want more privacy around your relationship with Dave that's fine," I say, "but I don't think it's fair to blame me for being curious. After all, you've always told me every detail of your private life before." Delicately, I hint that maybe Dave has something to do with the change, that perhaps he doesn't approve?

Evelyn sends back a blazing letter:

Yes you are correct in assuming Dave doesn't approve—not of what you are writing—but certainly the way you are treating me in this matter to which I most strongly agree with him. I have tried asking you to respect my wishes in this matter—which you obviously choose to ignore and continue in your bulldozing fashion—so sadly I now ask you to continue future discussions purely on a professional level.

. . .

One night I call Andrea and she can't talk. She calls back just as I'm about to put my kids to bed and I tell her that now *I* can't talk. It's obvious from her snippy response that I've pissed her off again. A few days later, she sends an e-mail:

This is not about my not accepting that you're busy. I get that you're very busy and that this is not a priority. My problem is with your lack of consideration for my feelings in this process. I think we should just take a break. And in response to your question about what I'm doing, the only thing happening in my life is my Dad, what's happening with him, and what it's bringing up for me.

I write back, rehashing the details of the telephone nonexchange: she was busy, then I was busy, my wife was working, my mother was visiting, my kids needed attention.

I tried to explain this to you and you took it as my being unwilling to make time for you. The irony is that I've been trying to make time, but I just can't do it on your schedule in the precise way that you demand. It seems there's nothing I can do that will please you.

After that I just don't contact her.

. . .

Then Evelyn writes again. In a tone of weary despair, she says that Jocelyn is having a terrible time and that they've all gone "way past rock bottom." She closes with this:

Please don't ask me to go into details because I won't, but I would sincerely like to apologize for that last e-mail. Dave wrote it and I stupidly signed it, something I will always regret. I won't make excuses but I would like to apologize for sending it. I can't take back what was written but I can say sorry. It is not what I want.

. . .

Then from Andrea:

Are you taking the break that I suggested a couple of e-mails ago? Have you given up on us? Are you overwhelmed? Are you out of town? Are you in rehab?

. . .

And Evelyn sends another bombshell, this one addressed to "dear everyone" and cc'd across the world. "I have decided to speak out," she announces. This has been the worst year of her life. No one will ever know the pain and sorrow she has experienced, the anguish of divorce and the torment of leaving her other children and how she felt abandoned by her so-called friends, who all took David's side and never once stopped to think how she felt:

None of you will know the anguish of having a child with a disability, none of you will know of the hardships associated with the issues that are a fact of life in such a situation. NONE OF YOU!!! I HAVE DONE NOTHING WRONG! I have received some hateful mail accusing me of abandoning my children. Well I have NOT abandoned them. I have taken ONE of them to get the surgery that she so richly deserves. I hope that

NONE of you EVER has to make the decision I had to make . . . EVER . . . because GOD help you if you do!!!

It goes on and on for ages. At the end of the letter, she announces her wedding to Dave. They've set the date for April.

. . .

Andrea's father dies. When I call to offer sympathy, she asks me if I'm taking notes.

. . .

Just before Jocelyn's cast comes off, David writes me from Australia. He says he was getting ready to leave for Baltimore so he could be there on the big day when he got an e-mail from Jocelyn telling him not to waste his time. He wasn't welcome. He called the hospital and Jocelyn was icy, accusing him of disowning her and refusing to pay the bills, leaving them "on the bread line," and when he tried to explain and even offered to show her receipts, she accused him of lying. It all sounded exactly like Evelyn's recent bombshell newsletter telling all her old friends to go to hell, he says. "Why is SHE cutting her friends away for??? Alecia is adamant that Dave is behind all this BUT WHY??? What is his motivation??"

. . .

Dave writes too, saying that he feels "somewhat responsible" for the distance that has come between Evelyn and me and wants to explain. Over the last twelve months, he says, they've been subjected to "one of the most hateful smear campaigns that anyone could possibly endure not just from Evelyn's own family but from friends and associates as far afield as Hawaii." All they want to do is see Jocelyn through her recovery, but lately it's been getting so hard they've been wondering if they could stick it out:

So I guess to some extent it has been my doing that we have put up shutters against the world so we can focus on seeing things through. To this end I truly apologize to you—sometimes in a battle like we have been through, there are innocent victims—and I guess that I am guilty of making you one of those. I am offering no excuses other than all Evelyn and I want is for a chance to be able to have the walks along the beach with Joc that we have dreamed of—just to be able to live like normal people with just our normal share of everyday problems.

. . .

There's a thaw with Andrea too. Who knows why? One night on the phone, I tease her about her hidden motivations. "You just like me because I'm mean. It's some kind of masochist thing."

She answers very seriously. "If you had been nice, I don't know if I would have trusted it."

Another time, I tell her about Jocelyn's operation and how I began to see the beauty in her.

"If you say that thing about we look wrong again, or even get close to it, it's going to threaten our friendship."

"But it's complicated," I say, telling her about my conversation with Dr. Kopits after the operation and how he disdained "cosmetic" work, and how he told me during that visit to his office that there are things in dwarfism so painful he has to hide them from himself. I think this is one of the things he's hiding, because it's something he can't cure.

"Yes," she says, "it's complicated. More than people want to admit." Her voice has gone cold again. "But there's an assumption that we do look wrong. It's not a shared conception. I don't get that from everybody. I get that from you."

"I accept that," I say. "There are people who have grace. Good people. I'm not one of them."

This is progress for me. Two years ago, I would have said they were in denial at best, hypocrites at worst. But Andrea exhales loudly. "You make it a moral thing. I don't think it's a moral thing for them. They just don't see it the way you do."

"You're right. And I admire them for it."

"It's not *admirable.* You're doing the moral thing again."

Before long, we're back in the same old fight.

. . .

Then I visit Evelyn and Jocelyn for the last time. It's cool and sunny in Baltimore that Saturday, the fresh smell of spring just around the corner. When I arrive at the hospital room, Jocelyn is lying flat on her back as usual. But the cast is gone. They cut it off two days ago. Her little legs are free and amazingly straight, and her cheeks are almost pink. There's even some gloss to her hair. She tells me the cast was really smelly when it came off. They celebrated with donuts and Italian food, and she's already doing physical therapy in the pool and starting to think about going back to school. She's exhausted but cheerful—more cheerful than I've ever seen her.

But it's unnerving trying to talk with Evelyn sitting two feet away, head down, typing furiously. She looked up when I came in the door and let me see her drawn, exhausted eyes, then went right back to typing.

The room is pure hospital this time. No Piglet pillowcases, no stuffed bears on the dresser. There's a feeling of grim purpose.

An hour later, Evelyn turns off the computer. Right away the phone rings and Evelyn picks it up. The man on the other end of the line starts yelling. It sounds like Dave, but I could be wrong and I'm sure as hell not going to ask. Evelyn listens for a minute and hangs up without saying a word and stares off like a prisoner of war.

"What happened?" I ask.

"I can't explain it," she says, numbly shaking her head. "Someday maybe I can tell you. Right now I just want it to go away."

We sit quietly for a minute or two, wrapped in the silence of the sickbed, just as we have so many hours and days before. I think about how all this started almost exactly two years ago, when I left my mother in another hospital and went to Atlanta to the dwarf convention. And one year later the same damn sickroom scene again, sitting there in my father's little room in Mexico, with him lying there gaunt and toothless and dying as I typed away on my laptop in the wing chair next to his bed.

Finally she begins to talk, her voice unbelievably weary. "I've had enough. I've had enough. I can't take it anymore. She depends on me for everything. She can brush her hair and that's it."

I just nod.

The hardest part, she says, has been the way people have acted.

"Your family?"

"I don't want to talk about it. Every time I talk about it just makes it worse. I just want it to be over."

Silence again. Evelyn has the trapped and hopeless look of a woman in a tower in one of those old English movies, imprisoned by an evil queen. She heaves a sigh. "Jocelyn's better and that's the important thing."

"Do you want to get a cup of coffee?" I say.

"I don't want to leave Jocelyn."

We sit in silence for another two minutes, maybe three, then I get up to kiss Jocelyn good-bye. I press my lips to her bulging forehead.

Evelyn looks up. Her eyes flash. "I don't regret anything I did," she says. "I did it for Jocelyn."

Fifteen

NINE MONTHS GO BY. Day after day, sitting at my desk, I try to make sense of my journey in and out of the Little World. What *is* the difference between Being and beings? How much does it matter? And once we've figured that out, how can we live with it a little better than we do? I don't come to any grand conclusions, but the patterns do become clearer. There are hospitals and pain, obsessive mothers and critical fathers, the fear of exposure and the need to be seen, a yearning to be normal and a corresponding swerve inward or away and all the mixed and dark feelings that swerve inspires. Eventually it comes down to love. That is the ultimate test that takes you down to the essence of what you are and the essence of what you hope to be. Love is the door where Michael trembles, where Meredith hesitates, where Andrea rages, where Jocelyn will someday learn the meaning of doubt. I don't exempt myself from any of this. I could have chosen other dwarfs to write about. I could have chosen Gibson Reynolds, who seemed so reasonable and well adjusted. There were others who offered themselves. But I was drawn to the ones who were difficult.

Then one insomniac midnight, reading in my office, I come across a case study of a woman who was born with a port wine stain that covered half her face. She was intelligent and attractive and a gifted artist, but she'd never had a relationship that lasted longer than two months. She was treated like a freak at school and had only one friend, a girl with warts covering her hands. At home, she was lost in a family of five children and felt she got attention only because of her flaw, but she became so dependent on her mother that she felt "fused" with her and dreamed of escaping. She frequently hid under the dining table. At sixteen she discovered makeup and that became a physical boundary between her and the world, easing the dread of meeting people. By her twenties she was so addicted to makeup that she felt she would "die from shame" if anyone ever saw her without it. "The makeup is a concrete manifestation of Tina's illusion that she can hide and make herself invisible," wrote her therapist, Maria T. Miloria.

For the first few years of treatment, Tina dodged Miloria's questions. She would come late, withdraw into herself, forget what she had said the week before. Miloria decided that she was suffering from avoidant personality disorder, which centers on a terrible fear of feeling exposed. But Tina didn't quit therapy. She kept showing up. She would hint at secrets she never quite told, make approaches to intimacy and then quickly withdraw. It was, Miloria decided, a kind of "hide and seek" behavior. "The dilemma of avoidant personalities can be conceptualized as follows," she wrote. "If they expose themselves to others (that is, seek), others will see them as defective and run away from them, and they will be abandoned. However, if they stay away (that is, hide), although they will be safe from the anticipated dread of being seen as defective, they will feel isolated and alone."

But Tina had one bit of hope—her painting. This is also common to people with avoidant personality disorder, Miloria wrote. "Often these individuals are creative, apparently trying to re-create themselves symbolically." The problem is that the creative impulse often develops into "grandiosity and exhibitionism and all kinds of magical fantasies." Usually the fantasy takes the form of standing in front of an admiring audience and hearing applause and gasps of admiration—of finally being seen as talented and beautiful.

Most of the time, this fantasy is just another source of heartbreak. But Tina actually was talented. Although she was tormented by grandiosity and self-loathing, seeing her work as either brilliant or horrible, she kept working. Little by little, her painting improved. She got shows and good reviews. Her confidence grew.

Then she ran out of money. Unable to pay for therapy, she said she wanted to take a break. But Miloria told her she thought things were going well and offered to treat her for free. That was the breakthrough moment. Tina realized that Miloria really cared about her and finally, after seven years, washed her face and let Miloria see her without makeup. From then on, she got better.

With all my heart, I wish I could tell a similar story here. The funny thing is, I see so much of myself in Tina—the grandiosity, the self-loathing, the fantasies of being admired. The fear of exposure, the need to be seen. On this level the gap between us—and Andrea and Michael and Kay and Martha and all those dwarfed in body or mind—is really so small. Surely we can reach across it.

But that's not how it happened. A few weeks after Jocelyn finally got cut out of her cast and flew back to Australia, I spent a day with Andrea in New York. The conversation was pleasant and sometimes warm, with a healthy amount of mutual ribbing. At the end of the day, we asked a man on the street to take our picture, realizing too late that he was severely retarded. I waited gamely while he fumbled with the camera, but Andrea stopped him and snatched it away (a bit brusquely, I thought) and gave it to the next guy who came along. As we posed, she leaned against me, resting her head on my shoulder. I told her I figured we were real friends now and likely to stay that way.

The next time we talked on the phone, she told me all her friends thought the story about the retarded man was funny as hell. I pretended to be scandalized. "Did you tell them how mean you were and how compassionate I was?"

"Actually, I was perceptive enough to realize that he might be uncomfortable."

She said it in her dry superior voice, and I couldn't help laughing at her.

"You know, what kills me is that you tell all your friends this totally warped version of things and they're going, 'Oh poor sweet Andrea.' "

"Nobody calls me sweet," she said.

When she hung up, I really thought we were in tune at last, buddies in black humor. And then it happened again, this time because she taped our photos to the wall of her office cubicle and I sent her one of my teasing notes: "You taped them to your wall? Hey, wait a sec—isn't that kind of like publishing them? Don't I get a say in this? What if people 'relate to me on the basis of what they've seen'?" She sent back a testy response and I told her she was being a pain in the ass again and she dodged a few phone calls and I missed a few more, and when we did finally hook up, she was angry all over again. And I got mad. How many more times was I going to have to apologize to this woman? Was it really a friendship or some kind of twisted psychodrama I was letting myself get dragged into out of sick reasons of my own? I spent a few days thinking it through and sent her what seemed to me a very grown-up, measured letter:

Having thought over this recent turn of events, I continue to feel somewhat baffled and frustrated and annoyed. Your reactions seem so blown out of proportion as to be symptomatic, a projection of your cathexis with the outside world or your dad or some damn thing. It's like you sit there and brood on minor things until you turn the ordinary things that happen into this big dark event.

After that there was another thaw and another flare-up and another ultimatum. "I thought you were going to call this week," she wrote one day. "I simply don't know how to do a friendship with you. I give up."

In response, I sent one of my stiffest notes so far:

Frankly, I was depressed and just wasn't in the mood to be dragged over the coals once again for something I don't think I did. If I could call and shoot the shit and have a few laughs like I do with my friends, then I would. But I'm always being forced to apologize and I just didn't fucking feel like it, okay?

We went through a few more rounds of this, and one night I found myself telling her that I was done with apologies, that I was never going to say the words "I'm sorry" to her again. "Anyway, that's not what you really want. What you really want is some kind of declaration of love—*that's* what this is really about. And even though I like you and admire you and you're tough and you're a fighter and a lot of other good things that I would like to have in my life, I refuse to be some kind of substitute for the Daddy Who Rejected You. It's not fair and it's not real."

Then she said she had an idea. "It's something I've been thinking about. Something that might salvage our friendship."

"Yeah? What?"

"I want us to go into therapy together," she says.

My first thought shames me still: A dwarf woman and a tall man in therapy together, trying to solve the riddle of difference—wouldn't that be great for the book!

My second reaction was a stab of dread. Over the last year, I'd come to understand that deep down in the heart of the fear we normals have for dwarfs is a subliminal intuition that they are the ultimate moral tar baby, sticky with our deepest feelings of fear and justice and truth and beauty, and if you touch them even lightly you might never get loose. And you can line up all the saints and postmodern philosophers who ever lived and it still won't change the truth, which is that this fear *isn't just prejudice.* Look at Dr. Kopits, sacrificing his family and career. Look at Emily the Tall Girl, dating dwarfs because her brother is a dwarf. Look at Evelyn. Ask why she took that defiant and suggestive and plaintive Internet name, Free.

Then I remembered what Andrea said to me about taking notes after her father died, and all the times she wondered if she was a real friend or just material, and I realized I was at one of those pivotal moral moments. Should I choose what's best for my book? Should I chuck the book and act like a friend? And what about that tickle of dread?

"Let's compromise," I said. "If this happens again, we'll go into therapy."

"No," she said, firm and implacable. "It *will* happen again."

I told her I'd think about it. That night, I explained the situation to my wife. She gave me a strange look, like she couldn't believe I was even considering the idea. "You don't go into therapy with a friend," she said.

A week later I sent Andrea this note, my last:

I'm sorry it has taken me so long to get back to you. I just didn't know what to do. I know that I can't go into therapy with you. That's just not what I call a friendship. I also know that it annoyed me when I suggested a compromise and you refused. You never want to compromise. You want to drag things out and torture me because I dared to teasingly call you a pain in the ass—or because I wrote something you didn't like or because I didn't want to talk on the phone or whatever. And if I don't go along with you, you issue ultimatums. Again, that's not what I call a friendship. So you're just going to have to do what you're going to have to do.

I haven't heard from her since.

Six months later, Michael called me up and started screaming at me. He'd seen a small notice in a magazine about the progress of this book. "You can't find anything else to write about? You can't leave this and let it die? I'll get the media involved in this! I'll let them know how I was deceived! The betrayal that was involved!" He calls me dumb. He calls me a jerk. He starts going on about "those ridiculous comments about big heads and big butts" and threatens to sue me.

I take it for about half an hour and then I start yelling back. "They *do* have big butts, goddamn it. I refuse to call them 'vertically challenged individuals.' "

"How can you be so heartless?" he says.

"Bullshit," I shout. "I'm not heartless, and you know it."

"And all that stuff that made me look like a pompous asshole—I'm beautiful, Meredith's beautiful, and the rest of you people are freaks. I said over and over to you, don't fuck me."

That's when I really lose it. "You fucked *me*, Michael. You and your pissy self-pity. I'm sick of taking shit from you. You say you want to be treated like everyone else and I made a strategic decision to do just that and paint the scene as vividly as I could just like I would do with politicians or movie stars or any other fucking person I would write about. And I knew that it would upset a few people, but I decided it was the right thing to do. I was trying not to *patronize* you. Trying to treat you just like grown-up people who can take it, and I've been beat up for it ever since by you and Meredith and every other fucking dwarf in the fucking Little World. You say you want the truth and you want to be treated like everyone else but you don't—you just want more happytalk bullshit about big hearts in little bodies and how we're all the same under the skin and beauty is only skin deep and I'm not going to do that! It's a fucking *lie!*"

To my amazement, he starts to cool down. He's still shouting, but with a significantly less hostile tone. "Why don't you put that in the book, about being patronized and big hearts in little bodies?"

"It *is* in the book."

It's as if a gale-force wind suddenly died. In a sad voice, he says he's dropped out of the LPA and yes he's irrational and yes he's angry and yes he's his own worst enemy. "But the LPA was the only place I ever felt important in my life, and the minute they read the descriptive words, which I understand were honest, once they read that, the rest of it was completely devalued. And that *exposed* me."

That word again.

"I know, Michael," I say. "I know how hard that is for you."

"I used to walk around kissing hands and babies," he says. "I was the mayor."

"I know you were."

And then he tells me that if I really want to tell the truth and blow up this "big hearts in little bodies" bullshit, then he has a lot to say. A lot to say. We spend another fifteen or twenty minutes repeating ourselves, going over the same old bloody ground, letting off steam. A month later, we meet in Los Angeles at the ultratrendy Mondrian Hotel, where all the women look like models and even the bellboys look like the idle rich. He doesn't want me to tape him, just to describe what I've been writing and what my thinking is, so I describe this book. And two hours later, he says he's feeling much more comfortable and we should start working on setting up some interviews. And there is some good stuff to tell. Meredith put her doctorate on hold to pursue an acting career and immediately landed one of the lead roles in a movie called *Unconditional Love.* She'll be starring opposite Kathy Bates. And she's got a good part in an upcoming episode of *NYPD Blue.* And they're getting married in the spring.

Then we say good-bye. As he leaves the hotel, I get that same wistful pang I feel every time I say good-bye to a dwarf and watch him walk away from me. Maybe it's a trick of perspective, the feeling that they are near but far at the same time, out of reach even before they're gone. He seems more burdened than he once did and perhaps a bit more sad, but he pulls himself together with an embattled dignity that touches me in a deep place—down in the place where we're all stuck together, struggling against terrible odds. I like him. I wish we could be friends.

But it doesn't work out that way. We exchange a lot of e-mail and a few phone calls, but we never manage to get as far as doing a proper interview. It's the old hide-and-seek routine. In one of his last notes, he says he's dropped into a deep depression and is back in therapy. "I'm no good to anyone now," he says. "I'll be in touch soon."

. . .

And Jocelyn? And Evelyn? When it came to the crunch, Jocelyn realized she still wasn't ready to leave her mother, so they flew off together to the west coast of Australia and a new life with Dave. I spoke to them just before they left for the airport, and Jocelyn seemed full of excitement and newfound happiness—giddily, she told me how she'd bought magnetic jewelry and she was going to stick it on her ears and nose and lips and totally shock *everyone* when she got off the plane. Evelyn was the complete opposite, in the same bleak mood as on that last day in the hospital when she was so miserable she seemed to want to disappear into the earth. And I knew that maybe the kindest thing would be just to let her.

But I couldn't. Hide-and-seek is my game too. So a few months later, I wrote notes to each of them asking how they were doing. Evelyn responded first, sounding like her old chipper self again. "While life is great, it is not idyllic!! It is definitely Australian outback. Gosh, red dirt everywhere—ugh!! I own more cleaning products than the supermarket right now." She and Dave were doing fine and yes, they closed each day with a walk on the beach, holding hands and watching the beautiful west Australian sunsets. And she still dreamed of moving to the United States someday and getting another taste of that intoxicating kindness.

Jocelyn was the one who surprised me. In my note, I asked her a bunch of questions about what was really going on in her mind these last two years. In response she sent back, as she proudly noted at the top, the longest letter she had ever written in her life, rambling giddily about getting out of her cast and the grueling months of physical therapy until she was finally strong enough to walk with crutches. "For the first time in my eighteen years of life nothing was blocking my way—NOTHING!!!! (OK, maybe something—like a chair, a bag and a step. Whoever invented the step, I really want to meet!)." She agreed that she did "knuckle down for a long time," but said she wasn't really trying to keep things under control like I thought. It was more just filtering out the small stuff. "It was only really the major problems that affected my life that I could deal with, and the others just had to take a number and wait their turn!" As to my question about the bond with her mother, yes, it was definitely tested. "The worst time was being stuck in hospital and being so dependent on everyone and hating it! The only thing that got me through was that I was going to walk and be VERY independent for the rest of my life!" She had big plans, starting with getting her driver's license and a set of pedal extensions and then going to college in the United States—by herself. And getting married and having children "and most definitely having an excellent quality of life."

As to her confidence, that was still coming:

Physically I am getting better everyday. In the next few weeks I should be moving onto canes and then you should soon just be able to eat my dust!!! I am definitely getting more vocal. Everyone now knows that I am here!!!

And somewhere in there, mixed into the burbling stream of words coming out of this once silent girl, is a phrase that still lingers in my mind, stuck on permanent replay because it came from so much pain and may still return to pain—will surely return to pain—yet is so full of hope. And hope like that, earned at such a cost, deserves respect. And maybe even a little hope in return.

This is what she said:

Oh the future—what a wonderful invention!

Andrea's name was changed at her request. For readability, some of the letters and dialogue were slightly edited or altered and the sequence of a few scenes moved.

Acknowledgments

THIS BOOK STARTED WITH A PHONE CALL from David Granger, the editor of *Esquire* magazine. Mark Warren guided it through the early stages. I can't overstate my debt to these two big-hearted men. They are serious and ambitious and work like dogs and for three years they've pushed me to do the best work I could do. Sandy Dijsktra saw the article as a book and sold it to HarperCollins, where David Hirshey and Jeff Kellogg did noble work trying to improve it. Heather Schroder gave me wise counsel. Kathy Potter never doubted me. Chris and Tanis Furst were supportive friends. Jennifer and Eleanore Richardson read carefully when it counted. Evelyn Powell I thank ten times for her generosity and vast courage, which included exposing much more of her inner life than she sometimes wanted to expose. More thanks to Jocelyn for never faltering; to Dr. Kopits for his time and wisdom and to Michael Gilden and Meredith Eaton for all they contributed, willingly and less willingly. I'm also grateful for the kindness and encouragement and naked honesty of Angela McTate, Gibson Reynolds, Martha Holland, Jennifer Arnold and all the other little and tall people who appear in this book. Without you I could not have done it, nor wanted to.

And to the woman I called Andrea, I still wish things were different.

About the Photographs

Page 1: Martha Holland, taking a break from the LPA ball to talk to the author.

Page 21: Jocelyn Powell, recovering from her last round of surgery. (Photo by Evelyn Powell)

Page 83: Angela McTate, holding her four-year-old son Joshua at the LPA convention.

Page 127: Michael Gilden and Meredith Eaton, in Santa Monica, California.

Page 141: Jocelyn and Evelyn Powell, at the Children's House in Baltimore just after her first round of surgery. (Photo by the author)

Page 163: Evelyn and Jocelyn Powell, arriving in Australia after the first round of surgery. (Photo by Judy Boyce)

Page 201: Gibson Reynolds and a friend, posing for a good-bye photo at the end of the convention.

Page 213: Dawn Lang, preparing for the LPA ball.

Page 249: Jocelyn Powell, the day before leaving Australia for her second round of operations. (Photo by the author)

Except as noted, all the photographs were shot by Miles Ladin.